Toxicology and Regulatory Process

Toxicology and Regulatory Process

Edited by

Sidney Green

Howard University College of Medicine
Washington, D.C., U.S.A.

CRC Press

Taylor & Francis Group

Boca Raton   London   New York

CRC Press is an imprint of the
Taylor & Francis Group, an **informa** business

CRC Press
Taylor & Francis Group
6000 Broken Sound Parkway NW, Suite 300
Boca Raton, FL 33487-2742

First issued in paperback 2019

© 2010 by Taylor & Francis Group, LLC
CRC Press is an imprint of Taylor & Francis Group, an Informa business

No claim to original U.S. Government works

ISBN-13: 978-0-8247-2385-9 (hbk)
ISBN-13: 978-0-367-39108-9 (pbk)

A CIP record for this book is available from the British Library.

Library of Congress Cataloging-in-Publication Data available on application

Visit the Taylor & Francis Web site at
http://www.taylorandfrancis.com

and the CRC Press Web site at
http://www.crcpress.com

Foreword

The nation needs better information to make evidence-based environmental health regulatory policies and to implement public health and treatment strategies to improve human health. This dearth of information has an enormous impact on the cost of health care and environmental regulatory compliance. The potential to develop new information and to translate toxicological discoveries into public health and clinical practice has improved remarkably over the past decade. For example, environmental health regulatory agencies are now debating the application of data derived from studies using genetically modified animal models and DNA and protein array technologies in human risk assessment. This exciting potential could have been realized sooner had the leadership of the biomedical research enterprise had a better appreciation of the important role that social and environmental factors play in human health and disease. In large measure, this failure in leadership and vision has contributed to the quagmire in which we now find ourselves with respect to our inability to blunt the rising epidemic of chronic diseases. This is unfortunate considering that we have known for more than a century that a connection exists between environmental exposures and disease risk. The toxicology community failed to make a compelling case for public investment in this area of research and technology development. However, over the past decade, the scope of the biomedical research enterprise has been enlarged to include social and environmental etiologic factors and population health. As a consequence, the discipline of toxicology is now experiencing a renaissance, fueled by the application of new cutting-edge technologies in genomics and proteomics, and is poised to have a major impact on both environmental health regulatory policy and the practice of medicine.

Historically, the foundation of environmental health risk assessment has been largely based on studies conducted at high doses and in animal models, which may not be highly relevant to safety assessment in humans. But, thanks to the development of technologies capable of monitoring effects of low-dose exposures at the level of gene and protein expression, structure, and function, one can now gain more mechanistic insights into toxicity and can better evaluate the predictability of surrogate models.

The National Institute of Environmental Health Sciences, with the advice and support of Dr. Sidney Green and other contributors to this publication, has led the way in promoting the new "omics" era of toxicology. Starting in 1997 and continuing through 2004, National Institute of Environmental Health Sciences has developed several new research initiatives to respond to the urgent need to understand the role of the environment in the development of chronic diseases. Such programs include the Environmental Genome Project in 1997, the institutionalization of the National Center for Toxicogenomics in 2000, and various mouse sequencing and model development projects over the past three years. Information and other resources resulting from these investments will provide generic information relevant to risk assessment and prevention and treatment of most chronic diseases.

Individuals vary, often significantly, in their response to environmental exposures. The variability provides a high "background noise" in studies to identify environmental links to disease. This variability often masks important environmental contributions to disease risk, and it is a major impediment in environmental medicine. Fortunately, the Human Genome Project led to the development of tools that can now be used to identify the genetic variations in the human genome that influence susceptibility to various environmental agents. The Environmental Genome Project was developed to catalog these genetic variants (polymorphisms) and determine their role in human susceptibility to various environmental agents.

The various mouse sequencing and model development projects were initiated to improve the relevance of toxicological testing in mouse models to humans. Although mouse studies can indicate the potential of an exposure to cause cancer or other diseases, there is no way to precisely extrapolate such studies to risk in humans. An important step in overcoming this uncertainty will be the identification of species differences (polymorphisms) between environmental response genes in humans and in various mouse strains. The National Institute of Environmental Health Sciences is now supporting a project to sequence the DNA of the 15 most commonly used mouse strains in toxicology, to pinpoint genetic differences that account for strain-specific differences in toxic response. The ultimate benefit of this sequencing effort will be to improve our ability to use mouse studies to determine the molecular basis of various human diseases.

Finally, National Institute of Environmental Health Sciences created the Comparative Mouse Genomics Centers Consortium Program to develop transgenic (knock-in and knock-out) models based on DNA sequence variants in environmental response genes discovered in the Environmental Genome Project. These genetically modified mice will serve as models to investigate the significance of human DNA polymorphisms in human health and disease.

In summary, the knowledge and technologies being developed through the above investments will provide risk assessors with critical information to predict how genetic background and interactions between multiple genes, proteins, and pathways influence human health and disease. Such studies are important for understanding interspecies and interindividual variation in response to environmental xenobiotics. Once we have validated the use of the new molecular approaches in predicting toxicity, determining how to use this information in risk assessment represents yet another challenge. The transition of toxicology from a descriptive discipline to a more mechanistic, predictive science has reinvigorated the field and its public health relevance in the era of chronic diseases.

This book fills a unique need of providing important information both for the general public and professionals involved in human risk assessment.

Kenneth Olden
Director, National Institute of Environmental Health
Sciences and the National Toxicology Program
National Institutes of Health
Research Triangle Park, North Carolina, U.S.A.

Preface

The aim of *Toxicology and Regulatory Process* is to acquaint the readers with the interrelationship between toxicology and the process of regulating chemicals/materials. At first blush this relationship may be regarded as regulation being the engine that drives toxicology. That is, toxicological methods do not just appear but are devised in response to regulatory needs. The obverse is also true: Toxicology drives the regulatory process. New methods in toxicology influence what approaches regulatory agencies take in solving issues regarding hazard determination, exposure, and risk assessment. This book will provide a greater depth of understanding about the relationship between these interdependent sciences and possibly will foster an atmosphere in which the mutually beneficial aspects of the relationship can be further enhanced.

Toxicology and Regulatory Process will not explore the many and varied standard toxicological methods used by regulatory agencies in their day-to-day activities. (That type of information is readily available elsewhere.) This book will delve into how and why certain toxicological methods or approaches evolved. For example, how and why did endocrine disruptors become a concern relative to adequate toxicological methods? The details associated with whether the science does or does not warrant such a concern would also be discussed. This would be an example of regulation driving toxicology. What influence has toxicogenomics had on regulating chemicals? This would be an example of toxicology driving the regulatory process. What are the lessons learned from these examples that can improve the relationship? Most important is the outcome of these efforts, that is, what is the state of their development today and possible reasons for that outcome?

This book consists of two major sections. One section provides information on how toxicology impacts the regulatory process, and the other section details information on how the regulatory process impacts toxico-logy. Each section provides examples and gives details regarding how and why the method/approach or legislation was developed, how it was intended to be used, and current views and status.

Toxicology and the Regulatory Process is not an academic textbook in the true sense, nor is it a desk reference for pharmacologists or toxicologists in industry or government. The book is for those who seek a broader under-standing of the interaction between the science of toxicology and the agen-cies that use toxicology.

We believe such a book will be of interest to pharmacologists and toxi-cologists in the industrial as well as governmental sectors. Academicians will also be interested, particularly those teaching and training individuals seek-ing employment in government and industry.

Sidney Green

Contents

SECTION II: INFLUENCE OF TOXICOLOGICAL RESEARCH ON THE REGULATORY PROCESS

Contributors

Cynthia A. Afshari Amgen Inc., Thousand Oaks, California, U.S.A.

Charles M. Auer U.S. Environmental Protection Agency, Office of
Pollution Prevention and Toxics, Washington, D.C., U.S.A.

David Brusick Bumpas, Virginia, U.S.A.

Mamata De Center for Drug Evaluation and Research,
U.S. Food and Drug Administration, Silver Spring, Maryland, U.S.A.

Alan M. Goldberg Department of Environmental Health Sciences,
Bloomberg School of Public Health, The Johns Hopkins University,
Baltimore, Maryland, U.S.A.

Richard H. Hefter[†] U.S. Environmental Protection Agency,
Office of Pollution Prevention and Toxics, Washington, D.C., U.S.A.

William L. Jordan U.S. Environmental Protection Agency,
Office of Pollution Prevention and Toxic Substances, Washington,
D.C., U.S.A.

John L. Kough Biopesticides and Pollution Prevention Division,
U.S. EPA, Office of Pesticide Programs, Washington, D.C., U.S.A.

[†] Deceased.

Barbara A. Leczynski   U.S. Environmental Protection Agency, Office of Pollution Prevention and Toxics, Washington, D.C., U.S.A.

Paul A. Locke   Department of Environmental Health Sciences, Bloomberg School of Public Health, The Johns Hopkins University, Baltimore, Maryland, U.S.A.

Owen McMaster   Center for Drug Evaluation and Research, U.S. Food and Drug Administration, Silver Spring, Maryland, U.S.A.

Christopher J. Portier   National Institute of Environmental Health Sciences, National Institutes of Health, Research Triangle Park, North Carolina, U.S.A.

William L. Roth   Office of Food Additive Safety, U.S. Food and Drug Administration, College Park, Maryland, U.S.A.

Harry Salem   United States Army, Edgewood Chemical Biological Center, Aberdeen Proving Ground, Maryland, U.S.A.

Wendelyn J. Schmidt   Center for Drug Evaluation and Research, U.S. Food and Drug Administration, Silver Spring, Maryland, U.S.A.

Gary E. Timm   U.S. Environmental Protection Agency, Office of Science Coordination and Policy, Washington, D.C., U.S.A.

Margaret Johnson Wilson   U.S. Environmental Protection Agency, Office of Pollution Prevention and Toxics, Washington, D.C., U.S.A.

Mary S. Wolfe   National Institute of Environmental Health Sciences, National Institutes of Health, Research Triangle Park, North Carolina, U.S.A.

Chris A. Wozniak   National Program Leader for Food Biotechnology and Microbiology, USDA-CSREES-PAS/CP, Washington, D.C., U.S.A.

1

The High Production Volume Challenge Program

Richard H. Hefter,[†] Barbara A. Leczynski, and Charles M. Auer[a]
*U.S. Environmental Protection Agency, Office of Pollution Prevention and Toxics,
Washington, D.C., U.S.A.*

INTRODUCTION

The U.S. Environmental Protection Agency's (EPA) High Production
Volume (HPV) Challenge Program is a prime example of how an alternative
approach to the traditional "command and control" rulemaking process can
help achieve impressive results and greater efficiency of resource utilization
by promoting involvement and cooperation among a diverse group of stakeholders to achieve critical public health and environmental objectives. Under
the Toxic Substances Control Act (TSCA) (15 U.S.C. 2601, et seq.), the U.S.
EPA is empowered to require the submission of data on the health and environmental effects of chemical substances that are subject to the Act and are
either manufactured in or imported into the United States. However, TSCA

[†] Deceased.

[a] The authors have held a variety of technical, policy, and managerial positions in EPA's Office
of Pollution Prevention and Toxics. Any views expressed in this chapter are personal to the
authors and are not necessarily the views and do not represent the policy of EPA. Any references to trade names, products, or programs of any non-EPA entity including public interest
organizations, trade associations, or companies is for informational purposes and does not constitute endorsement by EPA or by the authors. Any views expressed or text presented elsewhere
in this publication, apart from this chapter, are not those of these authors.

specifies that it is the responsibility of the industry to conduct testing, if needed, to develop this information. Traditionally, data have been developed through a rulemaking process, following procedures that are time consuming and resource intensive for both the Agency and the regulated community. Unfortunately, this process has often been adversarial in nature. By working in parallel with the traditional statutory-specified process of acquiring this information, the HPV Challenge Program has significantly accelerated the collection and public availability of baseline health and environmental effects data on more than 2100 widely used industrial chemicals. It has demonstrated the power of voluntary partnerships among the industry, environmental groups, and the EPA to achieve results in cases where a regulatory approach would likely have been much less effective.

The key to the tremendous success of the HPV Challenge Program, and what distinguishes it from past efforts to obtain information on the health and environmental effects of industrial chemicals through the traditional rulemaking process, is the cooperative and flexible nature of the program. The HPV Challenge Program achieved its success via a collaborative commitment by the EPA, the industry, and the environmental community to the fundamental objective of the program—to provide the U.S. public access to baseline health and environmental effects data so that they are informed about the potential hazards of the chemicals that are present in many of the products they use and to which they come in contact with in their daily lives. The program, from its inception, has been conducted in an open and transparent manner. Information submitted to the Agency is posted on the EPA website (1) for public viewing within 10 days of receipt by the Agency. Technical guidance documents prepared for the program are posted on the EPA website as draft documents so that public comments can be obtained. The documents are subsequently revised, as appropriate, in response to stakeholder input. The HPV Challenge Program has also been extremely innovative in its approach to data development, by promoting the use of categories of chemicals with similar properties and/or functionalities and Structure–Activity Relationships (SAR) to allow for data interpolation and extrapolation. These alternative approaches conserve resources in various ways including minimizing the use of animals for any new testing deemed appropriate. In fact, most of the guidelines, developed for use in the HPV Challenge Program, have been adopted by the international community for use in complementary health and environmental data collection programs, including the Organization for Economic Cooperation and Development (OECD) HPV Screening Information Data Set (SIDS) Program, after which the HPV Challenge Program was modeled (2).

The data that are obtained under the HPV Challenge Program serve multiple purposes. Availability of the data promotes the public's right-to-know about the potential hazards of chemicals that are found in their environment, their homes, their workplaces, and in the products they buy, and it enhances community health and environmental decision making. The data greatly improve government action, if warranted, on industrial chemicals by enabling the EPA and other federal, state, and local agencies to make informed decisions, based on sound science, to safeguard the public health and the environment. The data are also used to support and to improve voluntary industry product stewardship programs such as Responsible Care® [American Chemistry Council (ACC), Virginia, U.S.A.]. When combined with exposure and use information, these baseline data provide a framework for priority setting by the government and the industry, for future risk management actions, including evaluations for safe use under proper conditions. Empowered with this information, a range of beneficial responses will occur much earlier than they would otherwise. Examples of such actions include reduction and/or elimination of uses of chemicals that are identified as posing greater risks; substitution of safer chemicals for a range of uses; better understanding and management of all chemicals for which hazard information is now available to the public; more effective targeting of the EPA, state, and community resources across the various media for management of chemicals that pose the greatest risks; and better decision making by consumers on the use of these chemicals and the products containing them. Based on the EPA's experience with its pollution prevention programs, including the Design for the Environment Program (3), it is expected that the substitution of safer chemicals and the design of safer products and processes will also be accompanied by substantial savings for the industry. Furthermore, the use of safer chemicals in industrial processes and consumer products ensures that the goals of pollution prevention and environmental protection are being achieved.

The impetus for the HPV Challenge Program was a series of analyses conducted in the late 1990s, which indicated that little or no hazard data were publicly available on high-volume chemicals. To remedy the situation, the HPV Challenge Program set the ambitious goal of making publicly available by 2005 baseline health and environmental effects information on approximately 2800 HPV chemicals, i.e., those chemicals that are manufactured in or imported into the United States in quantities of one million pounds or more per year. As of July 2004, 401 companies, either individually or through 103 consortia, have voluntarily sponsored the screening of 2222 chemicals under the program. Screening-level hazard and physical–chemical properties data provided in 353 test plans are available on 1266 chemicals including 114 categories at the cited website (1). The test plans describe how data gaps will be filled and document how existing information submitted by industry sponsors meets internationally accepted, scientifically valid guidelines for data adequacy, thus ensuring that all data submitted by the industry under this program meet accepted quality standards. The HPV Challenge Program has set in motion a voluntary collaborative program, which by 2005 will produce an unprecedented increase in the public availability of baseline hazard data; more baseline hazard data will be available to the public as a result of this program than have been made available through traditional regulatory approaches since the TSCA went into effect in 1977.

BACKGROUND

The HPV Challenge Program was launched on October 9, 1998 by the EPA in association with the ACC, Environmental Defense (ED), and the American Petroleum Institute. Industry response to the "challenge" was from the beginning and continues to be extremely positive. Not only has the U.S. domestic chemical industry responded, but the international chemical producers community has also joined forces to complement the effort through the International Council of Chemical Associations (ICCA) (4). The chemicals sponsored for screening through the ICCA HPV Initiative are being submitted to the OECD HPV SIDS Program. Many chemicals are, in fact, sponsored under both programs.

The EPA believes that the U.S. public has a strong interest in and a fundamental right-to-know about potential chemical hazards. The HPV Challenge Program directly addresses an important and significant gap in knowledge on the part of the U.S. public about the potential hazards of chemicals that are routinely used in daily life. The EPA, ED [formerly the Environmental Defense Fund (EDF)], and the ACC [formerly the Chemical Manufacturers Association (CMA)] each conducted independent studies that found that baseline health and environmental effects information about HPV chemicals were generally not publicly available (5,6). The EPA's Chemical Hazard Data Availability Study (7) found that only 7% of HPV chemicals had publicly available data on all of the endpoints for which screening data are being collected under the HPV Challenge Program. Forty-three percent of the approximately 2800 HPV chemicals produced in the greatest quantity in the United States did not have any data available to the public regarding their baseline health and environmental effects.

The HPV Challenge Program is based on stakeholder involvement. In this way, the interests of the Agency's various stakeholders, including the chemical industry, environmental groups, animal welfare organizations, and the general public, played a significant role in the design and continue their participation in the implementation of the program. For example, the fundamental principles of the program worked out by the ACC, ED, and others prior to the initiation of the program are reflected in a "framework" document. The principles outlined in this document ensure that the goals of the industry, the environmental community, and the EPA are achieved cooperatively. They address timeframes for participation and submission of data and include a 2004 deadline for the completion of new testing and a 2005 deadline for all data to be made publicly available. They also set target percentages for the numbers of chemicals to be sponsored annually in the program so that back loading of submissions in the latter years would not occur. This scheme ensured that the public had access to this information in an ongoing and timely manner without undue burden on the industry to collect the information. Allowing categories of chemicals with similar characteristics to be sponsored through consortia (groups of companies working together to locate existing data and develop any new data needed) provides the industry with the incentive to share the costs and resources of making the data publicly available, ensuring that there is no duplication of testing, thus minimizing the number of animals used in any new testing conducted to fill data gaps. Making all information on the status of commitments publicly available and posting of data submitted by the industry within 10 days of receipt by the EPA on the Internet ensured that the data could be used as quickly as possible by all those interested.

Regular and frequent communication among a diverse set of stakeholders through a series of widely attended public stakeholder meetings in Washington, D.C., and in several EPA regional offices around the country raised awareness of the program and ensured that all stakeholders have a say in the evolution of the program. As new groups of stakeholders are recognized, the vehicles for communication and decision making about implementation of the program are expanded to accommodate these additional participants. The much praised Chemical Right-to-Know website (8), where all information is made publicly available, is updated regularly so that all interested parties are aware of new developments under the HPV Challenge Program. In addition, an e-mail update notification system, accessible through the website, provides automatic notification of updates and additions to the site to the program participants.

INTERNATIONAL EFFORTS TO INVESTIGATE HPV CHEMICALS

The OECD is a collaborative body of 30 member countries devoted to the development and refinement of economic and social policies. The OECD Environment Directorate (9) is responsible for a wide range of environmental issues related to sustainable development, transportation, agriculture, pollution prevention, and chemical safety, testing, and assessment.

In 1990, the member countries agreed to undertake the investigation of HPV chemicals in a cooperative way. These HPV chemicals include all chemicals reported as produced or imported at quantities greater than 1000 tons (2,200,000 pounds) per year in at least one member country or in the European Union region. The decision implies that member countries cooperatively

- select chemicals to be investigated,
- collect effects and exposure information from government and public sources and encourage the industry to provide information from their files,

- complete the agreed dossier for the SIDS by testing, as needed, and
- make an initial assessment of the potential hazard of each chemical investigated.

The overall objective of the OECD HPV Chemicals Program is to cooperatively undertake an initial assessment of HPV chemicals to screen them and agree on the need for further action. The program has undergone various changes in procedural as well as policy aspects, since its inception, to meet the needs of the national/regional programs of member countries. With this background and in light of the desire to significantly increase the output of the program and to make best use of various industry initiatives announced at the end of the 1990s, a major refocusing was agreed to in 1998. The overall HPV Chemicals Program was divided into six segments, each with distinct outputs and a clearly defined mechanism for oversight. The aim was to increase transparency, efficiency, and productivity of the program, and to allow for long-term planning by governments and the industry.

The refocused program comprises the following segments: (*i*) maintenance of the consolidated OECD HPV chemical list, (*ii*) improvement of tools to select chemicals from the HPV list for investigation, (*iii*) enhancement of the SIDS testing program, (*iv*) streamlining of SIDS initial assessments to focus on hazards, (*v*) coordination of post-SIDS work, and (*vi*) pilot projects on joint International Program on Chemical Safety (IPCS)/ OECD detailed international risk assessments.

Since 1999, the OECD has been concentrating on the first four segments—those related to selection, data gathering, testing, and initial hazard assessment. Detailed exposure information gathering and assessment is carried out in follow-up at the national (or regional) level when necessary but is no longer part of the SIDS initial assessment. Detailed international assessment of risks to human health and/or the environment is also no longer carried out under the guise of SIDS initial assessments but, rather, will be undertaken jointly by OECD and IPCS for appropriate pilot cases.

The content of the SIDS was adopted in November 1989 and revised in February 2000. The SIDS content, organized under five headings— substance information, physical–chemical properties, environmental fate, environmental toxicology, and mammalian toxicology—is presented on the facing page. Another change adopted by the OECD at this time, a direct result of efforts conducted under the HPV Challenge Program, was the use of robust study summary templates for data reporting. Robust summaries permit the submitter to make public the data that are summarized with sufficient detail to allow for the conduct of a screening-level assessment, yet not allow a third party to gain access to the raw data developed under the study. As such, robust summary data do not provide the level of detail that would be reported in a full study. Additional information on the content of the SIDS data set can be found at the cited website (2).

Substance Information

- Chemical identity

   Chemical Abstract Service Registry Number(s) (CASRN)
   Name [OECD names(s)]
   CAS descriptor (for inorganic chemicals)
   Composition of the chemical(s) being assessed

- Quantity (estimated production and/or import volume)
- Use pattern
- Sources of exposure

Physical–Chemical Properties

- Melting point/boiling point
- Relative density (for inorganic chemicals)
- Vapor pressure
- Partition coefficient of *n*-octanol/water
- Water solubility
- Dissociation constant
- Oxidation–reduction potential (for inorganic chemicals)

Environmental Fate

- Photodegradation
- Stability in water
- Transport and distribution between environmental compartments
- Aerobic biodegradability

Environmental Toxicology

- Acute toxicity to fish
- Acute toxicity to daphnia
- Acute toxicity to algae
- Chronic toxicity (as needed, based on physicochemical properties)
- Terrestrial toxicity (only if there is significant exposure to the terrestrial environment)

Mammalian Toxicology

- Acute toxicity
- Repeated dose toxicity
- Genetic toxicity (point mutations and chromosomal aberrations)

- Reproductive toxicity (fertility and developmental)
- Experience with human exposure (if available)

A description of the OECD work on HPV chemicals can be found at their website (10).

While the OECD HPV SIDS Program served as the model for the key components of the HPV Challenge Program, there are some major differences between the programs. First, under the SIDS Program, an initial hazard assessment is prepared and communicated with readily available exposure data from the sponsor country to determine if further work on that chemical(s) is warranted. This could be the conduct of additional testing and/or the preparation of an exposure and/or risk assessment. Under the HPV Challenge Program, only test plans and accompanying robust summaries are submitted. Screening and assessment activities will subsequently be performed by the Agency and by other stakeholders as appropriate (e.g., as a contribution to the OECD program). Second, except for "closed system intermediate" chemicals, which can qualify for reduced testing (repeated-dose and reproductive-toxicity testing are waived) if minimal or no exposure to that chemical can be documented, exposure information is not submitted unless the sponsor chooses to do so. Under the SIDS Program, the sponsor country is required to provide readily available production, exposure, and use data. Third, the work performed by the United States under the SIDS Program is an iterative process of review and comment by the Agency and the company sponsors. All documents submitted to the OECD for consideration are fully agreed upon by the Agency and stakeholders as presenting the U.S. position. However, in the HPV Challenge Program, sponsors submit test plans and robust summaries for a 120-day public comment period. It is entirely up to the company sponsors to accept or reject the comments they receive when revising their documents.

GUIDANCE DOCUMENTS

To assist sponsors in meeting their commitments to the HPV Challenge Program, the Agency prepared a number of guidance documents that were designed to maximize the use of existing information and to utilize various estimation techniques, where scientifically justified, so that any new testing proposed under the program would be minimal. This would reduce costs substantially and address animal welfare concerns by reducing the number of animals used in the HPV Challenge Program. The principal guidance documents prepared for the HPV Challenge Program are (*i*) searching for chemical information and data, (*ii*) using SAR, (*iii*) developing chemical categories, (*iv*) developing robust summaries; (*v*) assessing the adequacy of existing data, and (*vi*) fact sheet on animal welfare. Other guidance documents include "What to Test" and "Testing Closed System Intermediates." All guidance documents can be found at the cited website (11). The most heavily utilized guidance documents in the HPV Challenge Program are the SAR, category development, data adequacy, and robust summary documents. Although many of the guidance documents were developed from either existing sources or collective current thinking on a topic, the robust summary guidance document was strictly an outgrowth of the HPV Challenge Program.

The robust summary guidance document was developed to address three concerns raised by stakeholders during the development of the program. The first was the industry's reluctance to supply full copies of existing studies because of proprietary information concerns. They did not want other companies to be able to use their data to meet other regulatory requirements (mostly international). The second was a more practical concern, that of having stakeholders and the Agency potentially review a massive quantity of paper to determine if the existing data were adequate to address that endpoint. The third was a desire to have a uniform format for the information being submitted to fill the database faster and with greater accuracy. The solution was to develop a template for each of the SIDS endpoints that would capture the most important details of a study such that a technical reviewer could judge whether the underlying study was adequate to address the particular endpoint.

ROBUST SUMMARY TEMPLATES

Each robust summary template identifies the information items or data elements that should be reported in a robust summary for a particular SIDS endpoint. The information provided should be as complete as possible because robust summaries concern the critical study or studies on which the assessment of the adequacy of the data for each SIDS endpoint is based. It is generally expected that the most adequate, reliable, and relevant study for each SIDS endpoint (i.e., the "key study") will be identified and reported to the fullest level of the template. Studies considered inadequate should be clearly indicated and the reason(s) why the study is considered inadequate should be documented.

Each section of the template has two types of fields—controlled vocabulary and free text. The controlled vocabulary fields are identified by individual bullets and must be filled out, while the free text field under each section, entitled "remarks," requires the input of optional information. The remarks field can be used as a means to further explain the contents of a particular section, as is done in the "discussion" portion of a publication in a scientific journal. For example, under the "results" section, unexpected results could be further explained (e.g., results seen were due to the complex nature of the test substance, deviations from protocol, etc.). Further details on robust summary templates can be found at the cited website (12). An example template for repeated-dose toxicity is provided in Figure 1.

TEST SUBSTANCE

• Identity

⇒ Remarks field for Test Substance (Use for any pertinent, test substance-specific remarks.)

METHOD

• Method/guideline followed
• Test type
• GLP (Y/N)
• Year (study performed)
• Species
• Strain
• Route of administration; oral (gavage, drinking water, feed), dermal, inhalation (aerosol, vapor, gas, particulate), other
• Duration of test
• Doses/concentration levels
• Sex
• Exposure period
• Frequency of treatment
• Control group and treatment
• Post exposure observation period
• Statistical methods

⇒ Remarks field for Test Conditions. Detail and discuss any significant protocol deviations and detail differences from the guideline followed including the following as appropriate:

– Test Subjects

· Age at study initiation
· No. of animals per sex per dose

– Study Design

· Vehicle
· Satellite groups and reasons they were added
· Clinical observations performed and frequency (clinical pathology, functional observations, etc.)
· Organs examined at necropsy (macroscopic and microscopic)

Figure 1   Repeated-dose toxicity robust summary template.

CATEGORIES

An innovative component in the design of the HPV Challenge Program was the use of categories to develop and report data. As previously noted, categories are groups of chemicals with similar structures and/or functionalities. It was recognized early on in discussions with stakeholders that the program would be extremely expensive (both monetarily and in animal usage) and lengthy in terms of the time needed to develop new data if the sponsored

RESULTS

- NOAEL (NOEL)
- LOAEL (LOEL)
- Actual dose received by dose level by sex, if known
- Toxic response/effects by dose level
- Statistical results, as appropriate

⇒ Remarks field for Results. Describe additional information that may be needed to adequately assess data for reliability and use include the following if available. Provide at a minimum qualitative descriptions of elements where dose effect related observations were seen.

- Body weight
- Food/water consumption
- Description, severity, time of onset and duration of clinical signs
- Ophthalmologic findings incidence and severity
- Hematological findings incidence and severity
- Clinical biochemistry findings incidence and severity
- Mortality and time to death
- Gross pathology incidence and severity
- Organ weight changes
- Histopathology incidence and severity

CONCLUSIONS

⇒ Remarks field with the ability to identify source of comment, i.e. author and/or submitter

DATA QUALITY

- Reliabilities

⇒ Remarks field for Data Reliability

REFERENCES (Free Text)

OTHER

- Last changed (administrative field for updating)
- Order number for sorting (administrative field)

⇒ Remarks field for General Remarks [Note – Use for any other comments necessary for clarification.]

Figure 1 (*Continued*)

chemicals were addressed on a chemical-by-chemical basis. Although a few categories had been prepared for the OECD HPV SIDS Program at that time, they were rather simple in design and covered only a few chemicals. As such, there was little practical experience across the industry with regard to the use of category approaches to report data. To address this knowledge gap, the Agency developed guidance on how to properly construct categories to facilitate this alternative approach, and strongly encouraged company sponsors to consider the use of categories in their data development efforts. The category guidance document was well received and has been used extensively in the program, with over 80% of the chemicals submitted in categories.

The adoption by the OECD, in 1999/2000, of many of the HPV Challenge Program guidance documents for use in the SIDS Program was also important. The guidance documents adopted by the OECD in essentially the same format as they were prepared for the HPV Challenge Program are (*i*) Guidance for the Use of SAR in the HPV Chemicals Programme, (*ii*) Guidance on the Development and Use of Chemical Categories in the HPV Chemicals Programme, (*iii*) Guidance for Determining the Quality of Data for the SIDS Dossier, and (*iv*) Guidance on the Use of Robust Summaries in the HPV Chemicals Programme. All of these guidance documents were subsequently incorporated into the "Manual for Investigation of HPV Chemicals" in April 2003.

RESULTS OF THE HPV CHALLENGE PROGRAM

As noted above, one of the most significant results of the HPV Challenge Program has been the use of the category approach to address the SIDS endpoints. Of the 353 test plans submitted, 114 are category proposals covering 1027 chemicals. This represents approximately 81% of the chemicals addressed by the test plans. An innovative feature of the category proposals is that they cover three distinct types of categories. The first can be called the "traditional" category. A traditional category is a set of chemicals sharing a similar functional group and backbone, or covering a regular increase in chain length over the category. The second type of category is a "complex mixture family." These categories are characterized with chemicals from a complex mixture of similar chemical structures, e.g., long-chain fatty acids and tall oils. They are commonly structured around a manufacturing process by which different products are made, differing only in the relative proportions of the structurally similar mixture components. The third type of category is a "process stream." These categories are typically petroleum-related groupings, again based on a manufacturing process, except that the mixture is composed of structurally different materials, e.g., paraffins, aromatics, alicyclics, etc.

Each of the three types of categories submitted to the HPV Challenge Program rely on a supporting hypothesis or underlying theory of how the chemicals are related and how data reported for one chemical can be used to predict the toxicological responses of similar chemicals in that category. The supporting category hypothesis varies with the category type. "Traditional" categories are usually premised on predictable increases or decreases in toxicity manifested by the category members or, in the case of ecotoxicity, a change in the toxicity manifested as the physicochemical properties change, e.g., increasing log K_{ow} where the toxicity is likely to shift from acute to chronic toxicity. For "complex mixture families," the hypothesis is based on the assumption that because the complex mixtures are composed of similar components the toxicity will be similar for the individual products in the category. Finally, for "process streams," the hypothesis is usually that increasing or decreasing amounts of a particular component will determine the toxicity of the various category members, e.g., removal of the aromatic fraction as the petroleum stream is further processed will result in decreased toxicity.

While about 30% of the categories initially submitted by company sponsors needed additional information or restructuring to support the category hypothesis (most have been revised by sponsors), the significant result was that a large number of chemicals were able to be characterized with a limited amount of additional testing in a period significantly shorter than that would be required if a chemical-by-chemical approach was used in developing the needed data.

The EPA's approach to the assessment of categories under the HPV Challenge Program has two principal steps. The first is a review of the category test plan and supporting robust summaries. In this step, the EPA comments on the reasonableness of the category hypothesis and the adequacy of the supporting data and any testing proposed. The EPA's comments are sent to sponsors at the end of the 120-day comment period. Sponsors revise their submissions based on the public and EPA comments received.

The second step involves preparation and submission, by the sponsor(s), of the "final category analysis" per the category guidance following completion of any needed testing. The analysis is used by the sponsor to indicate that the underlying category hypothesis is reasonable and that existing and new test data can be either extrapolated or interpolated to address the untested category members or endpoints. The EPA's review at this step would be to agree with the analysis as presented or to advise the sponsor that the EPA did not believe the category "held" and thus additional testing was necessary and/or the category needed restructuring, either by subdivision of the category or treatment of the category members as single chemicals.

Published/Unpublished Data

As discussed earlier, the prime impetus for the HPV Challenge Program was the consistent finding in the series of reports that publicly available hazard information was lacking on a majority of the 2800 U.S. HPV chemicals. Following the EPA's guidance, sponsors identify existing data on HPV chemicals and submit these data in the form of robust summaries. To determine the source of the existing data (i.e., published or unpublished data), each test plan posted on the EPA website as of September 2003 was examined (236 test plans). Table 1 presents the results for the SIDS ecological and human health endpoints and selected physicochemical and environmental fate endpoints (some endpoints are routinely determined with estimation techniques).

Table 1 The Number and Percentage of Published and Unpublished Studies Reported for Selected SIDS Endpoints

Endpoint discipline	Endpoint	Published	Unpublished	Total
Health effects	Acute B oral	252 (28%)	651 (72%)	903
	Acute inhalation	92 (27%)	250 (73%)	342
	Acute dermal	70 (17%)	344 (83%)	414
	Repeat dose	419 (46%)	495 (54%)	914
	Genetox in vitro	658 (51%)	639 (49%)	1297
	Genetox in vivo	216 (51%)	211 (49%)	427
	Reproduction/ development	337 (52%)	314 (48%)	651
Environmental effects	Acute B fish	148 (24%)	473 (76%)	621
	Acute daphnid	97 (22%)	348 (78%)	445
	Acute algae	65 (21%)	239 (79%)	304
Environmental fate	Biodegradation	196 (31%)	442 (69%)	638
Physicochemical properties	Water solubility	220 (57%)	169 (43%)	389
	Vapor pressure	195 (55%)	162 (45%)	357
	Partition coefficient	131 (49%)	138 (51%)	269
	Boiling point	283 (68%)	136 (32%)	419
Grand total		3379 (40%)	5011 (60%)	8390

Abbreviation: SIDS, Screening Information Data Set.

As can be seen from the numbers of studies listed in Table 1, the HPV Challenge Program sponsors have made a significant amount of previously unpublished data public. This illustrates that the sponsors, either individually or through the many consortia participating in the program, have made a concerted effort to bring forth existing data. This effort has had a significant effect on the amount of new testing that sponsors have proposed. This result is also in keeping with the Agency's experience implementing the testing authorities under TSCA Section 4 where the industry has made every effort to bring forth existing data in response to proposed testing actions.

In addition, although Table 1 primarily addresses the SIDS endpoints, many sponsors have submitted summaries of all readily available data. These include such non-SIDS endpoints as skin and eye irritation, dermal sensitization, neurotoxicity, cancer, and metabolism/pharmacokinetic data. Many sponsors stated that they wanted to provide the most complete picture of the toxicological data available on the chemicals they manufacture.

HOW SIDS ENDPOINTS ARE ADDRESSED

Directly related to the above discussion is an analysis of how sponsors have proposed to address the SIDS endpoints. The EPA analyzed test plans posted on the HPV Challenge website as of September 2003 to determine

Table 2 Test Plan Proposed Approaches to Fill Data Gaps for Ecotoxicity and Environmental Fate Endpoints[a]

Endpoint	Tests proposed	Estimated/SAR/ read-across	Submitted data
Acute fish	5.9	32.1	62.0
Acute daphnia	6.8	31.6	61.6
Acute algae	7.8	40.9	51.3
Partition coefficient	8.1	35.5	56.4
Biodegradation	6.8	30.8	62.4
Overall	6.8	34.9	58.3

[a] Reported as percentages.
Abbreviation: SAR, Structure–Activity Relationships.

how the health and environmental effects endpoints were addressed in the test plan. Three approaches to meet the minimum data requirements for each SIDS endpoint were examined. Data needs were met by (*i*) the use of existing scientifically adequate data, (*ii*) the use of an estimation technique such as SAR[a] or "read-across"[b] analysis or by giving a rationale for no testing, or (*iii*) the proposal of new testing. For human health effects, five endpoints (acute, repeat dose, reproductive, developmental, and genetic toxicity) were considered. For environmental effects, three endpoints (acute toxicity to fish, daphnia, and algae) were examined. The 236 test plans examined address 1113 chemicals in 97 categories and 139 single chemicals.

As can be seen in Tables 2 and 3, HPV Challenge sponsors have made maximum use of the guidance concerning the use of SAR and category proposals, and in combination with the significant amount of previously unpublished data made available through the program, only a relatively small amount of testing has been proposed. Overall, for health and environmental effects, less than 10% of the endpoints are proposed to be addressed with new testing. The exact mix and number of new tests may change as sponsors consider EPA and public comments and as the remainder of test plans are submitted; however, there is no reason to believe that the overall conclusions will change significantly. Comments could lead to some additional tests being performed but could also result in the need for fewer tests.

---

[a] SAR is the relationship of the molecular structure of a chemical with a physicochemical property, environmental fate attribute, and/or specific effect on human health or an environmental species. These correlations may be qualitative (simple SAR) or quantitative (quantitative SAR, or QSAR).

[b] Read-across is a data estimation approach used for category chemicals. It represents the intent to estimate a value (quantitatively or qualitatively) for an untested endpoint for a category member given the available experimental data on tested endpoints for other category members. A read-across value may be either interpolated (more confidence) or extrapolated (less confidence).

Table 3 Test Plan Proposed Approaches to Fill Data Gaps for Human Health Endpoints[a]

Endpoint	Tests proposed	Estimated/SAR/ read-across	Submitted data
Acute	2.0	21.8	76.2
Genetox[b]	4.7	35.9	59.4
Repeated dose	4.7	36.3	59.0
Reproductive	8.0	45.4	46.6
Development	8.2	45.0	46.8
Overall	6.2	44.1	49.7

[a]Reported as percentages.
[b]Genetox covers gene mutations and chromosomal aberrations.
Abbreviation: SAR, Structure–Activity Relationships.

Increased Understanding of the Hazards of HPV Chemicals

With the main impetus for the HPV Challenge Program being the various studies documenting that for a majority of the HPV chemicals used in the United States, few or no health or environmental effects data were publicly available, the fact that a significant amount of previously unpublished data are now available to the public serves to document the program's tremendous success. This achievement resulted from an unprecedented sharing of data among members of many panels and consortia formed to address the data needs of the chemicals they jointly manufacture. Until the launch of this program, this information resided in the files of the individual manufacturers. Consequently, for the first time, at the close of the program in 2005, SIDS-level data will be available on most of the HPV chemicals produced in the United States. In addition, because many companies submitted existing data beyond the SIDS endpoints, the public will have a much better understanding of the overall hazards posed by the chemicals to which they are exposed.

Next Steps

Given the unqualified success of the HPV Challenge Program to date, the EPA and a number of stakeholder groups are working on the next course of action for the use of the tremendous amount of data now being made publicly available. A critical step in this process is the availability of an electronic database to hold the data repository and to allow users to access and query the information to meet their needs. The EPA is currently developing an Oracle-based system, the HPV Information System (HPVIS), that will provide data access and retrieval capabilities not currently available on the EPA website. This system was scheduled for release in 2005. The EPA is also contacting companies that did not sponsor the HPV chemicals that they manufacture to encourage them to sign up with the program and ensure that all unsponsored (orphan) chemicals will have complete SIDS data sets.

The data development efforts described thus far rely on the voluntary participation of the global chemical industry in concert with governments and nongovernmental organizations to ensure that the data collected and made publicly available meet international standards. In announcing the Agency's HPV Initiative (5,6), the EPA indicated that data needs that remain unmet in the HPV Challenge Program or through international efforts may be addressed through TSCA rulemaking. The EPA further stated that any associated rulemaking will generally be carried out in a manner consistent with the OECD HPV SIDS Program to ensure that the data and information generated can be contributed to the international efforts and, conversely, that international SIDS testing and assessments can be used to fulfill the data gaps identified as part of the HPV Challenge Program or in related TSCA section 4 HPV SIDS rulemaking, thus avoiding unnecessary or duplicative testing and its associated costs. On December 26, 2000, the EPA issued a proposed rulemaking that addressed the unmet needs of 37 chemicals (13,14). The EPA is currently working on finalizing the proposed rulemaking and exploring options for meeting the remaining unmet data needs for the HPV chemicals used in the United States. Additional information on the HPV Challenge Program, including a description of the strategy to address the unmet data needs (orphans strategy), can be found on the EPA website (16).

The EPA recently established the National Pollution Prevention and Toxics Advisory Committee (NPPTAC). The NPPTAC provides information and recommendations on the overall policy and operation of programs managed by the Office of Pollution Prevention and Toxics, in performing its duties and responsibilities under the TSCA and the Pollution Prevention Act (15). In particular, the objectives of the Committee are to provide policy advice and recommendations in such areas as assessment and management of chemical risk, pollution prevention and risk communication, and opportunities for coordination.

Recognizing the important contributions made by stakeholders in the HPV Challenge Program, the NPPTAC has also formed an HPV workgroup to discuss various aspects of the HPV Challenge Program and its next steps. The HPV workgroup will initially focus on how to prioritize or screen the HPV data to determine which chemicals or categories of chemicals are of interest for additional data collection, including information on exposure and use scenarios and on hazard communication, assessment, or risk management. The group also plans to consider the HPVIS, orphan chemicals, and other issues related to the HPV Challenge Program. By engaging customers who will use the data for different purposes, much will be learned about how best to select chemicals for further investigation, thus ensuring that the public is aware of the hazards and risks they face when exposed to chemicals and products in their everyday lives.

CONCLUSION

The HPV Challenge Program set in motion a fundamental change in the way basic toxicological information is shared with the public. The data published in scientific journals and reported to governmental agencies are typically limited to studies that indicate a hazard or toxicological effect and thus suggest potential risks to exposed populations. The technical guidance that was developed for the program and shared with the international community provided the incentive for companies to report all existing data (both positive and negative results) for HPV chemicals and to organize related chemicals into categories. These data can now be used for several valuable purposes: to screen HPV chemicals using a consistent data set to identify higher (and lower) priorities for further action by EPA or others; to further develop and strengthen SAR models for estimation of effects across chemicals of similar physicochemical properties; and to apply the HPV Challenge data and the improved SAR models in identifying and developing lower hazard or less persistent new or existing chemicals that can substitute for higher hazard/more persistent chemicals. From these results, it is clear that the HPV Challenge Program has set the stage for a new paradigm in the understanding of the toxicology of industrial chemicals.

Through the voluntary collaboration of U.S. chemical manufacturers, environmental organizations, and the EPA, as well as the recognition by the OECD, its member countries, and the global chemical industry that the burden for developing and sharing data is an international responsibility, a wealth of data that can be used for many purposes by many stakeholders is now being made publicly available at a pace never seen before. This is a tremendous accomplishment that the public is clearly benefiting from now and will continue to do so in the years to come. These data provide the public with a better understanding of the hazards that may be posed by the existing HPV chemicals. The ability to readily access and effectively apply this information is helping to ensure that new, more environmentally friendly chemicals and processes are designed. Given the complexities of the current regulatory requirements, an achievement of this scale and speed would not have been possible without the voluntary HPV Challenge Program.

REFERENCES

1. www.epa.gov/chemrtk/hpvrstp.htm
2. http://cs3-hq.oecd.org/scripts/hpv/
3. www.epa.gov/opptintr/dfe

4. www.icca-chem.org
5. Federal Register 2000; 65(248):81,686–81,698.
6. www.epa.gov/chemrtk/ts42213.htm
7. www.epa.gov/chemrtk/hazchem.htm
8. www.epa.gov/chemrtk
9. www.oecd.org/environment
10. www.oecd.org/document/21
11. www.epa.gov/chemrtk/guidocs.htm
12. www.epa.gov/chemrtk/robsumgd.htm
13. Federal Register 2000; 65(248):81,658–81,685.
14. www.epa.gov/oppt/chemtest/hpv.pdf
15. www.epa.gov/oppt/npptac
16. www.epa.gov/chemrtk/hpvreport.pdf

2

Homeland Security and Bioterrorism

Harry Salem

*United States Army, Edgewood Chemical Biological Center,
Aberdeen Proving Ground, Maryland, U.S.A.*

INTRODUCTION

Many acts of terror have occurred in the United States and elsewhere in the world throughout history (1–10). It is the events that took place on September 11, 2001 in the United States that has heightened the concern with acts of terror, and has resulted in President Bush's declaration of war on terrorism. The rest of the world has been encouraged to join this war to combat and eradicate terrorism globally.

DEFINITIONS

Terrorism has been defined by the Federal Bureau of Investigation (FBI) as the unlawful use of force or violence against persons or property to intimidate or coerce a government, civilian population, or any segment thereof, in furtherance of political or social objectives (11). The term "terrorism" originated from the eighteenth century French word "terrorisme" (meaning "under the terror") based on the Latin verbs "terrere" (meaning "to tremble") and "deterrere" (meaning "to frighten from") (12).

The use of weapons of mass destruction (WMD) by terrorists is a reality. WMD include chemical, biological, radiological, and nuclear (CBRN) agents as well as explosives. In addition to the classical, chemical, and biological warfare agents, there are toxic industrial chemicals (TICs), which have a definite chemical structure, and toxic industrial materials (TIMs), which do not consist of an exact or constant composition of a particular chemical, but a variety or mixture of many chemicals. Examples include asbestos, which is a composition of a variety of fibers and minerals, and gasoline whose composition varies with the octane level, manufacturer, and the season of manufacture. These include both TICs and replicating biological agents that can also be used as a more likely choice by terrorists (13).

Terrorist acts can be carried out by individuals, groups, or governments. Terrorism has been further subdivided into specific categories, examples of which are described below.

Bioterrorism

Bioterrorism is the act of an intentional release of a naturally occurring or human-modified toxin or biological agent. These agents are usually dispersed as an aerosol spray and are inhaled. To be effective against a large number of people, the aerosol particle size needs to be optimal. In addition, for the biological agents, there is an incubation period of several days before any symptoms appear (14). These same agents can be used to contaminate our food and water supply, and can be effective not only by inhalation, but also orally and cutaneously. Ashford et al. (15) defines bioterrorism as the intentional use of microorganisms or toxins derived from living organisms to cause death or disease in humans, animals, or plants on which we depend.

Domestic Terrorism

The commission of terrorist attacks in a state, by forces inside or originating from that state, is considered domestic terrorism, and is usually unexpected (16). More recently, the term is being used in any country or state in which the act occurs, whether the terrorist forces are internal or external. Examples include the World Trade Center bombings in 1993 and 2001, the Oklahoma city bombing, the pipe bombing during the Atlanta Olympics, the sarin attack in the Tokyo subway, and the numerous suicide, or more correctly described, homicide bombings in Israel.

Narcoterrorism

The term "narcoterrorism" was coined by the former President Belaunde Terry of Peru in 1983, in describing terrorist type attacks that were staged against his nation's antinarcotics police. Narcoterrorism was understood to mean the attempts or acts of narcotic traffickers to influence the policies of the government, the enforcement of the law, and the administration of justice, by systematic threat or use of violence. Probably, the best-known example of narcoterrorism is Pablo Escobar's ruthless dealings with the Colombian government. The term is increasingly being used to describe the activities of known terrorist organizations that engage in drug trafficking to fund their operations, and gain recruits and expertise. Some of these organizations include the FARC, ELN, and AUC in Columbia, Hezbollah in Lebanon, and Al-Qaeda throughout the Middle East, Europe, and Central Asia (17).

Cyberterrorism

The term "cyberterrorism" describes a largely hypothetical form of terrorism involving the use of cracking techniques over computer networks, in the place of direct physical attacks. As the Internet continues to expand, computer systems continue to be assigned increasingly greater responsibility while at the same time becoming increasingly complex and interdependent Thus sabotage or terrorism via cyberspace is probably inevitable and is a serious threat (18). Our nation is now fully dependent upon cybersystems (hardware and software) for the functioning of the government and economy and for critical functions such as transportation, electrical power, manufacturing, and communications; yet our networks remain vulnerable to relatively simple cyber attacks.

HISTORICAL

On September 11, 2001, two hijacked airliners crashed into the twin towers of the World Trade Center in New York City. Shortly after, the Pentagon was hit by another hijacked plane. A fourth hijacked plane crashed into a field in southern Pennsylvania, near Shanksville, after the passengers fought back the hijackers. It was suspected that this plane was bound for a high profile target in Washington, such as the Capitol or the White House. It was noted that the 19 hijackers were all young Arab males, who had their religion, ethnicity, and appearance in common (19). The attacks killed 3025 people including U.S. citizens and other nationals. President Bush and his Cabinet officials indicated that Osama Bin Laden was the prime suspect and that they considered the United States in a state of war with international terrorism. In the aftermath of the attacks, the United States formed the Global Coalition against terrorism (20,21).

Immediately following the attack on United States on September 11, 2001, the Department of Defense began combat air patrols over U.S. cities, and the Department of Transportation grounded all U.S. private aircraft. In addition, the Federal Emergency Management Agency (FEMA) activated the Federal Response Plan, and the U.S. Customs went to Level 1 Alert at all border ports of entry. For the first time ever, the Health and Human Services (HHS) activated the National Disaster Medical System, dispatching more than 300 medical and mortuary personnel as well as emergency push packages of medical supplies to the New York and Washington, D.C. areas. They also put 80 Disaster Medical Assistance Teams nationwide and 700 private sector medical professionals on deployment alert. In addition, the Nuclear Regulatory Commission advised all nuclear power plants, nonpower reactors, nuclear fuel facilities and gaseous diffusion plants to go to the highest level of security, which they did. The President ordered federal disaster funding for New York and FEMA deployed the National Urban Search and Rescue Response Team and the U.S. Army Corps of Engineers to assist in the debris removal. On September 12, 2001, FEMA deployed emergency medical and mortuary teams to New York and Washington, D.C. The Federal Aviation Agency allowed the limited reopening of the nation's commercial airspace system, thereby allowing the diverted flights from the previous day to continue to their original destinations. On September 13, the Department of Justice and Treasury deployed marshals, border patrol, and customs officials to provide a larger police presence at airports as they reopened. The following day, September 14, is when the President proclaimed a national emergency and ordered the ready reserves of the armed forces to active duty. On the same day, the FBI released the list of 19 suspected terrorists. On September 17, the Attorney General directed the establishment of 94 units of Antiterrorism Task Forces, one for each U.S. Attorney Office. On September 18, the President signed the authorization for the Use of Military Force bill and for additional disaster funding for New York. In his address to Congress on September 20, the President announced the creation of the Office of Homeland Security (OHS), and appointed Governor Tom Ridge of Pennsylvania as the Director. Of the five billion dollars the President released for disaster relief, more than 126 million dollars was announced for the support of health services for the attack by HHS on September 21. On October 8, the President swore-in Governor Ridge as Assistant to the President for Homeland Security, and issued the executive order creating the OHS. General (Retired) Wayne Downing was sworn-in as the Director of the Office of Combating Terrorism (OCT), and issued an executive order creating the OCT on October 9. The executive order, establishing the President's Critical Infrastructure Protection Board to coordinate and have cognizance of federal efforts and programs that relate to protection of information systems, was issued by the President on October 16.

Exactly one week after the attacks on September 11, 2001, five postal covers containing anthrax are thought to have been mailed to media outlets mostly in the New York area. The media outlets were ABC News, CBS News, NBC News, and the New York Post. The fifth letter was addressed to The National Enquirer's old mailing address and later forwarded to their new offices at American Media Inc. (AMI) in Florida. The Sun paper, also published at AMI, is where one of the anthrax victims, Robert Stevens, died. A note in the New York Post letter stated, "09–11–01, this is next, take penacilin now, death to America, death to Israel, Allah is great." These five letters are believed to have been mailed from Trenton, New Jersey on September 18, 2001. The anthrax in those letters appeared to be of the type that could have caused only skin infections or cutaneous anthrax, and not death. However, a second batch of anthrax letters mailed three weeks later on October 9, 2001, also from Trenton, New Jersey, contained a modern "weaponized" form of anthrax, previously unseen by bioweapons experts. These anthrax letters were addressed to two Democratic Senators in Washington, D.C. (Senator Patrick Leahy D-Vermont and Senate Majority Leader Tom Daschle) and both had a note that stated, "09–11–01, you cannot stop us. We have this anthrax. You die now. Are you afraid? Death to America. Death to Israel. Allah is great." Twenty-two people developed anthrax infections, mostly cutaneous. Five persons died of inhalation anthrax. In addition to Robert Stevens who died in Florida, it is believed that two individuals died from cross-contamination of the mail. Apparently, the weaponized anthrax from the Senate letters penetrated the porous envelopes and contaminated nearby mail, causing the death of Ottilie Lundren, an elderly woman from Oxford, Connecticut, and Kathy Nguyen, a Vietnamese immigrant from New York City. The other two deaths occurred among the employees of the Brentwood mail facility in Washington, D.C., Thomas Morris Jr. and Joseph Curseen. It appeared that they were exposed to the contaminated Senate letters as they went through the system. These incidents resulted in thousands of people in the United States taking the two-month course of antibiotics (Cipro) in an effort to preempt anthrax infections. All of the letters contained the Ames strain, although there were at least two different grades of anthrax. The political impact of these events was greater compared to the physical impact. Thus, attention was focused on biowar and bioterrorism, as well as on the promotion of biosecurity. The U.S. Food and Drug Administration (FDA) toughened regulations to protect the U.S. food supply of imported food from both accidental and human-introduced toxic substances. In addition, research to identify genetically modified bacteria with toxic genes was initiated in an effort to identify deliberate versus accidental attack more quickly. To date no one has been charged with these attacks (22).

Although the impetus and focus for the war on terrorism was heightened after the events of September 11, 2001, the U.S. Army Edgewood Chemical Biological Center (ECBC) and its predecessor organizations have been providing chemical and biological solutions to the armed forces since 1917, and has been applying the dual-use solutions to the civilian sector since at least 1996. In 1996, protecting the U.S. homeland was added to its mission. Since that time, ECBC has trained over 28,000 first responders in 105 communities across the United States as part of the Congressional Domestic Preparedness Program (23). Most of us working in chemical and biological defense have always anticipated terrorist acts using WMD against Americans. In 1998, the Secretary of Defense William Cohen verbalized this threat when he stated, "bioterrorism is a matter of not if, but when." We were again cautioned on February 13, 2001, when President Bush, in an address at the Norfolk Naval Air Station in Norfolk, Virginia, stated, "the grave threat from nuclear, biological, and chemical weapons has not gone away with the Cold War. It has evolved into many separate threats, some of them harder to see and harder to answer. And the adversaries seeking these tools of terror are less predictable, more diverse." Further, following the horrendous physical and deadly acts of September 11, 2001, and the subsequent dissemination of anthrax spores in the mail in the United States, Secretary Donald Rumsfeld testified at the Defense Subcommittee of the Senate Appropriations Committee on May 21, 2002. At that time he stated, "We do face additional terrorist threats. And the issue is not if, but when and where. And we need to face that."

The weapons we need to consider are those that can be used against civilian and military populations, as well as against our infrastructure and way of life. These include CBRN, explosives, and combinations of these. The chemical and biological agent spectrum is presented in Table 1. This table lists the clearly chemical, the clearly biological, as well as the midspectrum agents. In addition to these agents of military interest, we now also include TICs and TIMs. The clearly chemical agents include chemical warfare agents (blood, vesicant, choking, nerve, psychological, incapacitating, and riot control agents). The emerging chemicals that are protection defeating and physical incapacitants, as well as the nontraditional agents, are considered midspectrum agents. In addition, fourth generation agents cannot be ignored. The clearly biological agents include the pathogens (bacteria,

Table 1 CB Warfare Agent Threat Spectrum

CW		BW		
Clearly chemical		Clearly biological		
Classic chemical	*Emerging chemical*[a]	*Bioregulator*[a]	*Toxin*[a]	*Pathogen*
Blood	Protection defeating	Pain	Plant	Bacteria
Vesicant	Physical incapacitant	Sleep	Bacterial	Viruses
Nerve	NTA	Blood pressure	Venom	Rickettsiae
Psychological incapacitant (BZ)		Mood enhancers	Marine	Genetic engineered microorganisms
Choking			Fungal	
Toxic industrial chemicals			Algal	

[a]Midspectrum.
Abbreviations: BW, biological warfare; CW, chemical warfare; NTA, nontraditional agents.

viruses, rickettsia, and genetically engineered microorganisms), while the mid-spectrum biological agents include toxins, bioregulators, or physiologically active compounds. The potential types of novel or engineered agents include engineered organisms for enhanced aerosol and environmental stability, converting benign organisms to produce toxin or venom subfractions or endogenous bioregulators, or to immunologically alter microorganisms to defeat standard detection, identification, and diagnostic methods, or to cause microorganisms to become resistant to antibiotics, standard vaccines, and therapeutics. Combinations of these can be used with improved delivery systems.

The Center for Disease Control and Prevention (CDC) has further categorized biological threat agents as A, B, and C (24). Although these agents are considered inhalation threats, they can also contaminate buildings and crops. Those agents that are soluble, stable, and resistant to water treatment and disinfection can be considered a threat to our water supply. Category A includes organisms that pose a risk to national security because they can easily be disseminated or transmitted from person to person and cause high mortality rates. These agents have the potential for major public health impact and might cause public panic and social disruption. In addition, these agents require special action for public health preparedness. The Category A biological diseases and agents include anthrax (*Bacillus anthracis*), botulism (*Clostridium botulinum*), plague (*Yersinia pestis*), smallpox (*Variola major*), tularemia (*Francisella tularensis*), and viral hemorrhagic fevers [filoviruses (e.g., Ebola, Marburg), and arenaviruses (e.g., Lassa, Machupo)]. Category B includes organisms that are second priority agents which are moderately easy to disseminate and cause moderate morbidity and low mortality rates. These agents require specific enhancement of CDC's diagnostic capacity and enhanced disease surveillance. The diseases and agents include brucellosis (*Brucella*), epsilon toxin of *Clostridium perfringens*, those causing food safety threats (*Salmonella* species, *Escherichia coli* 0157:H7, and *Shigella*), glanders (*Burkholderia mallei*), psittacosis (*Chlamydia psittaci*), Q fever (*Coxiella burnetii*), ricin toxin from *Ricinus communis* (castor beans), staphylococcal enterotoxin B, typhus fever (*Rickettsia prowazekii*), viral encephalitis (alphaviruses such as Venezuelan equine encephalitis, eastern equine encephalitis, western equine encephalitis), and those causing water safety threats (e.g., *Vibrio cholerae, Cryptosporidium parvum*).

Category C includes organisms that are the third highest priority. These agents are the emerging pathogens that can be engineered for mass dissemination in the future because of availability and ease of production and dissemination. They have the potential for causing high morbidity and mortality rates. The Category C emerging infectious disease threats such as the Nipah virus and Hantavirus pose a major health impact (24). Although the CDC Category B agents include the water safety threat such as *V. cholerae* and *C. parvum*, Burrows and Renner (25) provide information on biological agents that need to be considered when evaluating potable water for the armed forces. In their publication, they include replicating agents and biotoxins that are either considered as definite water threats or considered probable, possible, or unlikely water threats, and report on whether they are known or unknown and probably or possibly weaponized, along with other characteristics. Table 2 lists the known replicating agents and biotoxin water threats to military potable water supplies, and indications as to which have been weaponized as reported by Burrows and Renner (25). Weaponized chemical and biological warfare agents usually refer to those agents that have been treated or processed to enhance their abilities for dissemination and toxicity or infection by the inhalation and dermal routes of exposure. A more generic definition of weaponized agents would be those that have been produced for dissemination in sufficient quantities to cause the desired effect by any route of exposure. The desired effect is scenario and mission dependent, and may be illness, death, or denial of life-sustaining drinking water or real estate.

The previous discussion relating to the potential chemical and biological threats against our troops in wartime and against our population and infrastructure are those considered effective by inhalation exposures via gas, vapor, or aerosol dissemination. These types of dissemination can result in respiratory and cutaneous exposures. In addition, respiratory deposition of agents trapped in the mucociliary escalator can move up and be swallowed, thus causing oral or gastrointestinal exposure. These same threats may or may not be applicable when it comes to contamination of our food and water supplies. To be an effective contaminant of our water supplies, factors such as solubility, stability, breakdown products, resistance to water treatment, and even weaponization must be considered. Some of these factors are discussed by Burrows and Renner (25). Contamination of our water supplies can result in human exposures via the oral, cutaneous, and inhalation routes. For example, drinking or washing and cooking food in the water can result in oral exposures, while bathing and showering can cause both cutaneous and inhalation exposures.

The literature reports that deliberate chemical and biological contamination of water supplies has been common in history. These have ranged from crude dumping of human and animal cadavers into water supplies to well-orchestrated contamination with anthrax and cholera. The contamination of wells, reservoirs, and other water sources meant for armies and civilian populations, with crude use of filth, cadavers, animal carcasses, and contagion, dates back to antiquity and continues even today (2). In 600 B.C.E., the Assyrians poisoned water wells with rye ergot, and Solon of Athens poisoned drinking water with hellebore roots during the siege of Krissa. The Cirrhaens drank the water and developed violent and uncontrollable diarrhea, and were quickly defeated (26–30). During the Peloponnesian War in 430 B.C.E., when plague broke out in Athens, the Spartans were accused of poisoning most of the water sources in

Table 2 Examples of Potential Bioterrorism Threats

Agent	Incubation period	Contagious	Signs and symptoms	Treatment
Bacterial diseases				
Anthrax	1–6 days	No	Flu-like chills, fever, swollen lymph nodes	Antibiotic vaccine
Plague				
Pneumonic	2–3 days	Yes	Chills, high fever, headache, spitting up blood, shortness of breath	Antibiotic vaccine
Bubonic	2–10 days	No		
Viral diseases				
Smallpox	7–17 days	Yes	Fever, rigors, pustules, vomiting, headache	Vaccine supportive therapy
Viral hemorrhagic fevers	4–21 days	Yes	Fever, vomiting, diarrhea, mottled/blotchy skin	Vaccine symptomatic treatment
Toxins				
Botulinum toxin (neurotoxin)	1–5 days	No	Weakness, dizziness, dry mouth and throat, blurred vision, paralysis	Supportive care, antitoxins, and vaccines
Ricin (cytotoxin)				
Ingestion	4–8 hr	No	Nausea, vomiting, bloody diarrhea, GI cramps, shortness of breath	Supportive oxygenation and hydration
Inhalation	12–24 hr			

Abbreviation: GI, gastrointestinal.

Athens (27). Persian, Greek, and Roman literature from around 300 B.C.E. quote examples of dead animals used to contaminate wells and other sources of water (31). About 200 B.C.E., the Carthaginians spiked wine with mandrake root to sedate the enemy (29,30). For thousands of years, cyanide has been used as a waterborne poison. In ancient Rome, around 50 C.E., Nero eliminated his enemies with cherry laurel water, which contained hydrocyanic acid (4). During the battle of Tortona in 1155 C.E., Barbaross used bodies of dead and decomposing soldiers to poison the enemy's water supply (29–31). During the American Civil War in 1863, the Confederate troops under General Johnston contaminated wells and ponds with animal carcasses during their retreat from Vicksburg. General Sherman and his troops, in pursuit of the Confederates, found the water en route unfit to drink (1,26,29,30). In the Second World War, the Japanese attacked at least 11 Chinese cities, intending to contaminate food and water supplies with anthrax, cholera, and various other bacteria. Hitler's forces released sewage into a Bohemian reservoir, deliberately sickening the population. In deliberate violation of the Geneva Convention, Yugoslav federal forces or those allied with them appear to have poisoned wells throughout Kosovo in October and November of 1998. They dumped animal carcasses and hazardous materials such as paints, oil, and gasoline in most of the area wells, deliberately sickening the populace and denying them the use of the wells (3,4). Earlier in 1939 and 1940, over 1000 Manchurian wells were laced with typhoid bacteria that were effective killers. Outbreaks of typhoid fever devastated many villages. About forty 13- to 15-year-old youngsters who drank lemonade made from the contaminated well water contracted typhoid fever and subsequently died (1). In 1970, the Weather Underground, a revolutionary movement, planned to put incapacitating chemical and biological warfare agents into urban water supplies to demonstrate the impotence of the federal government (5). A group of college students influenced by ecoterrorist ideology and the 1960s drug culture wanted to kill off most of humanity to prevent the destruction of nature and start the human race all over again with a select few. Their plan to use aerosol attacks and contaminate urban water supplies with eight microbial pathogens, including agents of typhoid fever, diphtheria, dysentery, and meningitis, was aborted on the discovery of these agents (5). In 1980, the Marxist revolutionary group called the Red Army Faction had cultivated botulinum toxins in a safe house in Paris, which was discovered along with plans to poison water supplies in Germany (5). In spite of this historical background, various governmental agencies have conducted discordant assessments, and have failed to evaluate the risk of sabotage contaminating water supplies with chemical and biological warfare agents. The President's Commission on Critical Infrastructure Protection (PCCIP) indicated that most water supply systems in the United States are vulnerable to terrorist attacks through the distribution networks (32). Senator Bill Frist, the Senate Majority Leader, and the Senate's only physician, stated that our water supply is generally considered safe, but that there is anxiety over the threat of bioterrorism on our water supply (33). The President's Commission concluded that the water supplied to U.S. communities is potentially vulnerable to terrorist attack by insertion of biological agents, chemical agents, or toxins. The possibility of attack is of considerable concern, and the agents could be a threat if they were inserted at critical points in the system, theoretically causing a large number of casualties (32). Bill Brackin of the Alabama Department of the Environmental Management Laboratory stated that after September 11, 2001, concern about safety of the nation's drinking water has increased. He described a variety of chemical agents that could be placed in a community's water supply. These included nerve agents, pesticides, cyanide, and radioactive materials (34). Nonfissionable radioactive materials such as cesium-137, strontium-90, and cobalt-60, which cannot be used in nuclear weapons, can be used to contaminate water supplies, business centers, government facilities, or transportation networks. While these radiological attacks are unlikely to cause significant casualties, they could still cause damage to property and the environment, and produce societal and political disruption (35). In April 1985, a threat was made to poison the water supply of New York City with plutonium. To evaluate the credibility of the threat, the New York City water supply was monitored for radioactivity. A sample taken on April 17 indicated a radioactive level of 21 femtocuries per liter, which was a factor 100 times greater than previously observed. This level, however, was substantially lower than the maximum of 5000 femtocuries considered safe for drinking water, under federal guidelines (36,37).

More recently, evidence has surfaced that terrorists trained in Afghanistan were planning to poison water supplies with cyanide, which would be easy to do and very hard to prevent (38). In January 2002, a bulletin from the FBI's National Infrastructure Protection Center pointed out that members of Al-Qaeda were trying to gain control of U.S. water supplies and wastewater treatment plants. Later, in July 2002, the FBI issued a series of warnings regarding possible terrorist attacks against American targets. The nation's water utilities were told to prepare to defend against attacks on pumping stations and pipes that serve the cities and suburbs. This action was a result of documents discovered in Afghanistan, which indicated that Al-Qaeda terrorists were investigating ways to disrupt the U.S. water supply on a massive scale (39). Again, in September 2003, the FBI issued a warning of the potential use of nicotine and solanine by terrorists to poison the food and water supplies (40).

It has been recognized that a greater threat to our water supply might exist from more commonly available chemicals and materials such as TICs and TIMs as well as naturally occurring pathogens (41). The nation's largest outbreak of a waterborne disease occurred in Milwaukee, Wisconsin, in the spring of 1993. *Cryptosporidium*, a protozoan, passed undetected through two water treatment plants. It caused over 400,000 illnesses (mostly diarrhea) and between 50 and 100 deaths in the 800,000 customers who drank the water. Although the cause was not clear, it confirmed that pathogens introduced in water supplies could cause death (39). The Marine Corps has confirmed and suggested that terrorists might employ TICs because they are easily available and accessible. In addition, recent attacks by terrorists have shown their preference to use assets of the host country, rather than smuggling in raw materials or weapon systems (13). While aerosol dispersion of an agent may be more effective in contaminating a larger number of people than through contamination of our water supplies, the PCCIP considered most water supply systems vulnerable to terrorist attacks primarily through the distribution network (32). Ideal poisoning agents that can be used to contaminate water should be reasonably or highly toxic. They should be highly soluble or miscible in water, have no taste or odor, and be chemically and physically stable and resistant to aqueous hydrolysis and water treatment. In addition, ideal contaminants should be difficult to recognize and detect, as well as have delayed activity and no known antidote (42).

It is believed that for the first time, in 1982, domestic chemical terrorism found its way into homes in the United States (43). Seven relatively young people in the Chicago, Illinois area collapsed and died after taking Tylenol capsules that had been laced with 65 to 100 mg cyanide per capsule. The lethal dose of cyanide is approximately 0.5 to 1.0 mg/kg, or about 70 mg for an adult person (44). These were the first victims to die from product tampering. The police believed that the murderer bought or stole the products from the stores, tampered with them, and returned them to the stores. The police speculated that the terrorist could have had a grudge against the producers of Tylenol, society in general, or even the stores where the tainted products were found and may have resided in the same area. The perpetrator was never apprehended even though Johnson and Johnson, maker of the capsules, offered a $100,000 reward (45). As a result of these events, in 1983, Congress passed the Federal Anti-Tampering Act. This legislation made it a federal crime to tamper with food, drugs, cosmetics, and other consumer products. In addition, many manufacturers have made their products tamper resistant (46). Copycat tampering followed involving Lipton Cup-a-Soup and Excedrin in 1986, Tylenol once again in 1986, Sudafed in 1991, and Goodies Headache Powder in 1992, all resulting in deaths. Sporadic tampering has continued in the Chicago vicinity, Detroit, and Tennessee (45).

It has been reported that during the World Trade Center's first bombing in New York City in 1993, sodium cyanide was present in the explosive and had been burnt in the heat of the explosion. It was speculated that had it vaporized, the cyanide gas would have been dispersed into the North Tower and tens of thousands would have been killed, whereas only six people died as it did not vaporize (47). More recently, it was reported by Hammer (48) that a Palestinian homicide bomber sprinkled an anticoagulant rat poison among the components of the bomb that he exploded in a bus in Jerusalem. Among the survivors, a 14-year-old girl was bleeding uncontrollably from every one of her puncture wounds. Using a coagulant drug eventually stopped the young victim's bleeding.

The poisonous anthrax letter campaign following the catastrophe of September 11, 2001 is also not novel. It has been reported that anthrax was used as a weapon by the Czech resistance in World War II against the German occupiers by spreading spores on envelopes (49). Letters containing anthrax spores were received in Senator Daschle's office in the Hart Senate building, and elsewhere in the United States. The CDC confirmed that five people exposed died of inhalation anthrax, six others contracted inhalation anthrax, and seven contracted cutaneous anthrax (50,51). In 1993, the Aum Shinrikyo, a Japanese religious doomsday sect, cultured anthrax and sprayed it from a rooftop for 24 hours in a Tokyo suburb. Although there were complaints about the smell, none of the neighbors reported any anthrax symptoms. It was reported that the anthrax was of the Sterne strain used in animal vaccines, which lacks the DNA fragment necessary to cause disease (52).

Throughout history, chemical and biological weapons (CBW) have been used by politically motivated individuals or groups for purposes such as assassinations and to terrorize. The terrorist threat to the United States was recognized even before the 1990s, and in January 1999, President Bill Clinton stated that United States would be subject to a terrorist attack involving CBW within the next few years (53). With the collapse of the Cold War, during which the Soviet Union attempted to expand communism and suppress religious expression, ethnic and cultural conflicts arose from many smaller regional states. Thus, it appeared that expansionist militant Islamic fundamentalism was replacing expansionist communism. The exponential rise in militant Islamic terrorism in the 1980s and 1990s, directed against U.S. targets, served to confirm that Iran would attempt to export its revolution throughout the Gulf Region and the Middle East, thereby threatening U.S. strategic interests such as the oil supplies and the stability of Israel (53). These regional states sought WMD in pursuit of their regional ambitions and transnational threats involving terrorism, crime, and drugs. Many of these radical regimes in the developing world are profoundly anti-American and are actively seeking to limit U.S. influence in their regions. Thus, they sponsor terrorism as a means of pursuing their foreign policy goals. The societal vulnerabilities to terrorism were heightened in February 1993 by the bombing of the World Trade Center in New York City by militant Islamic fundamentalists (53). This was followed by the April 1995 bombing of the Alfred P. Murrah Federal Building in Oklahoma City by American far-right extremists. There have, however, been allegations by an Oklahoma City reporter, Jana Davis, of an Iraqi connection with the Oklahoma City bombing and the catastrophic events of September 11, 2001 (54).

In Japan, the Aum Shinrikyo group used the chemical warfare agent sarin against civilians. On June 27, 1994, the group targeted a dormitory in Matsumoto, where three judges who had ruled against them in a land deal trial, lived. The group attempted to deliver the deadly sarin gas in the open. The chemical reaction went wrong and the wind changed direction. The judges survived, but became ill. Seven victims in the neighborhood died that evening, and over 200 were injured. The group's next attempt was to disseminate the sarin gas in a closed system. They chose the subway system in Japan and attacked on March 20, 1995, at 8:15 A.M. Members of the group pierced plastic bags containing 30% sarin with sharp tipped umbrellas and let the sarin evaporate into a lethal gas. Eleven passengers were killed and over 5500 were injured (55). Of these, over 4000 were considered primarily the "worried well."

Target sites for terrorists can be multiple and varied, and include anything imaginable as well as the unimaginable. The most common targets for contamination include everyday requirements for sustenance of life and activities, such as air, food, and water. Targets also include our infrastructure, transportation, historic landmarks, sports, and entertainment events. The Jihad manual that is distributed at Al-Qaeda terrorist training camps describes strategies for visiting holy wars on Europe and the United States. These include picking sites with high human intensity, such as football stadiums or skyscrapers, and symbols of great sentimental value, such as the Eiffel Tower and the Statue of Liberty. It is also stated in the manual not to settle for one strike at a time, but that four targets must be hit simultaneously... so that the government of the nation being attacked will know that they are serious. The attempt to target the Eiffel Tower on Christmas Eve in 1994 is also described by Miller et al. (8). Four Algerian terrorists from the armed Islamic group with ties to Osama bin Laden attempted to direct a hijacked plane into the Eiffel Tower. They held a gun to the head of an Air France pilot on Flight 8969, scheduled to fly from Algeria to Paris, and directed that he fly the plane into the Eiffel Tower. While the plane was even on the ground three passengers were killed, and the plane took off loaded with fuel and dynamite. Somehow, the crew convinced the terrorists that they needed to land at Marseilles for refueling. At the airport in Marseilles, the police stormed the plane and arrested the terrorists, thus averting the destruction of the Eiffel Tower and the loss of many more lives (56).

Terrorist activity against the citizens of United States was perpetrated in 1995 in Saudi Arabia in a bombing that killed 5 U.S. military personnel, as well as in 1996 when the bombing of the Khobar Tower in Saudi Arabia killed 19 and injured 200 U.S. military personnel. In 1998, terrorist bombing of the U.S. embassies in Africa killed 224 people and injured another 5000, and in the year 2000, the bombing of the USS Cole killed 17 and injured 39 U.S. sailors. In the terrorist attacks that took place on September 11, 2001, at the World Trade Center, the Pentagon, and in the downed fourth hijacked plane, an estimated 2800 U.S. citizens were killed.

Examples of bioterrorism are described below. In 1972, a fascist group, the Order of the Rising Sun, was caught with 80 pounds of typhoid bacteria cultures. Their plans were to poison the water supplies in major midwestern cities in the United States (52,57). In Paris in 1980, the German Red Army safe house was raided where the Baader-Meinhof gang of Germany was discovered to have *C. botulinum* cultures along with publications on bacterial infections. Their plans were to poison the water supplies in 20 German towns if radical lawyers were not permitted to defend an imprisoned comrade. Another incident occurred in the United States in 1984. An Indian guru, Bhagwan Shree Rajneesh, poisoned local salad bars in Antelope, Oregon, where 715 residents were affected by salmonella poisoning. This was done to influence the outcome of a local election (52,57). On April 24, 1997, a petri dish labeled "anthrachs" arrived in a package mailed to the offices of B'Nai Brith in Washington, D.C. The dish contained a red gelatinous material, which nine hours later was determined to be harmless. This was the nation's first major anthrax hoax, but not the last. The following year, reports of anthrax hoaxes were surpassing the phoned-in bomb threats (58–61). As a result, many states passed laws making hoaxes more serious than previously.

More than 20 countries are working towards a chemical warfare capability, as chemical warfare agents are relatively cheap and readily available. These include the blister, nerve, blood, and choking agents. It is estimated that more than 10 countries are developing biological warfare capability, and at least 20 countries have delivery capabilities or are developing them. In his State of the Union speech, President George Bush in 2002 declared that North Korea, Iran, and Iraq constitute the "Axis of Evil." In a speech entitled "Beyond the Axis of Evil," U.S. Undersecretary of State John Bolton included Cuba, Libya, and Syria, along with North Korea, Iran, and Iraq, as rogue states actively attempting to develop WMD (62).

The Dudley Knox Library of the Naval Postgraduate School Index of Terrorist Groups lists approximately 100 terrorist organizations and provides profiles on them. Most prominent on the list are the militant Islamic fundamentalist terrorist groups. These include Al-Aqsa Biyad, Al-Fatah, Al-Qaeda, Hamas, Hezbollah, Popular Front for the Liberation of Palestine, Palestine Liberation Organization, and the Islamic Jihad (63). We are currently living in a WMD age. Al-Qaeda, responsible for the devastating events of September 11, 2001, is not a state-based enemy, but is a worldwide terrorist network of cells operating in many nations. In addition to Al-Qaeda and other militant Islamic sleeper cells within the United States, there is also the danger posed by ecoterrorists. The FBI defines ecoterrorism as the "use or threatened use of violence of a criminal nature against innocent people or against property by an environmentally oriented, subnational group for environmental-political reasons." James F. Jarboe, Domestic Terrorism Section Chief, Counterterrorism Division of the FBI, in his statement to a congressional subcommittee hearing on environmental terrorism on February 12, 2002 stated that we have witnessed dramatic changes in the nature of the terrorist threat. During the past several years, special interest extremisms as characterized by the Animal Liberation Front and the Earth Liberation Front have emerged as serious terrorist threats. These groups feel justified in taking violent steps to protect animals and the natural environment from further harm (64).

President George W. Bush in his State of the Union address on January 29, 2002 addressed the issue of how we will respond to protect our population and infrastructure from acts of terrorism, and prevent such acts as well. He stated that "the United States of America will not permit the world's most dangerous regimes to threaten us with the world's most destructive weapons. . . . My budget nearly doubles funding for a sustained strategy of homeland security, focused on four key areas: bioterrorism, emergency response, airport and border security, and improved intelligence. We will develop vaccines to fight anthrax and other deadly diseases. We'll increase funding to help states and communities train and equip our heroic police and firefighters."

HOMELAND SECURITY

In a letter from the White House to his fellow citizens, President George W. Bush (July 16, 2002) announced that on October 8, 2001, he had established the Office of Homeland Security (OHS) within the White House and, as its first responsibility, directed it to produce the first National Strategy for Homeland Security. This strategy, a product of more than eight months of intense consultation with thousands of people, including governors, mayors, state legislators, members of Congress, concerned citizens, foreign leaders, professors, soldiers, firefighters, police officers, doctors, scientists, airline pilots, farmers, business leaders, civic activists, journalists, veterans, and victims of terrorism and their families, is a national strategy, not a federal strategy. As a national strategy, it requires compatible, mutually supporting state, local, and private-sector strategies. The national strategy creates a comprehensive plan for using talents and resources in the United States to enhance our protection and reduce the vulnerability to terrorist attacks based on the principles of cooperation and partnership. This strategy provides a framework for the contributions that we can all make to secure our homeland, and will be adjusted and amended over time, and will align the resources of the federal budget directly to the task of securing the homeland. The strategic objectives are (*i*) to prevent terrorist attacks within the United States, (*ii*) reduce the vulnerability of the United States to terrorism, and (*iii*) to minimize the damage from attacks that do occur and to aid quick recovery. The national strategy aligns and focuses homeland security functions into six critical mission areas that are as follows: (*i*) intelligence and warning, (*ii*) border and transportation security, (*iii*) domestic counterterrorism, (*iv*) protecting critical infrastructure, (*v*) defending against catastrophic terrorism, and (*vi*) emergency preparedness and response. The first three mission areas focus primarily on preventing terrorist attacks, the next two, on reducing our vulnerabilities, and the final one on minimizing the damage from attacks that do occur and on quick recovery.

Terrorist select their targets strategically and deliberately based on the weaknesses and vulnerabilities they observe in our defenses and preparedness. They can use WMD or conventional weapons to wreak unprecedented damage on United States. This could result in large numbers of casualties, mass psychological disruption, contamination, and significant economic damage, and overwhelm local medical capabilities.

Since virtually every community in the United States is connected to the global transportation network by the seaports, airports, highways, pipelines, railroads, and waterways that move people and goods within and out of the nation, we must protect the flow of people, goods, and services across our borders. We must also prevent terrorists from using these systems to deliver their implements of destruction. This concern is addressed in the national strategy by identifying six major initiatives in this area. They are (*i*) ensure accountability in border and transportation security, (*ii*) create "smart borders," (*iii*) increase security of international shipping containers, (*iv*) implement the Aviation and Transportation Security Act of 2001, (*v*) recapitalize the U.S. Coast Guard, and (*vi*) reform immigration services. The President also proposed to Congress that the principal border and transportation security agencies, the Immigration and Naturalization Service, the U.S. Customs Service, the U.S. Coast Guard, the Animal and Plant Health Inspection Service, and the Transportation Security Agency be transferred to the Department of Homeland Security to assist in implementation of these initiatives.

The President's proposal would also consolidate the artificial separation of "crisis management" and "consequence management" into incident management under the Department of Homeland Security in cooperation with state and local governments.

Including federal, state, and local law enforcement, the United States spends approximately $100 billion per year on homeland security.

The national strategy indicates that the nation's advantages in science and technology is a key to securing the homeland and concludes that just as in World War II and the Cold War, our strength in science and technology will bring us victory in the war on terrorism (65).

PATRIOT LEGISLATION ACT

The Patriot Legislation Act (HR 3162, S. 1510, Public Law 107–56) passed by a House vote of 357 to 66, and the Senate by 98 to 1, and was immediately signed into law by President Bush on October 26, 2001 (66,67). This law enhances the surveillance capabilities of the government by increasing its ability to conduct electronic surveillance, tracking finances, and requesting DNA information. It allows the arrest of those involved with money laundering, reorganizes some priorities in regard to immigration, authorizes rewards for citizens who help combat terrorism, and provides for an individual to monitor civil rights abuses (68). This law, Uniting and Strengthening America by Providing Appropriate Tools Required to Intercept and Obstruct Terrorism Act of 2001 (Patriot Act), was enacted in response to the September 11, 2001 terrorist attacks. Although the law may result in some constraints on civil liberties, our most basic civil right is the right to live without perpetual fear.

BIOSHIELD

In his 2003 State of the Union address, President Bush announced Project BioShield, a comprehensive effort to develop and make available modern, effective drugs and vaccines to protect against attacks by biological and chemical weapons or other dangerous pathogens. This will assure that resources, estimated to be 5.6 billion dollars over 10 years, are available to pay for next generation medical countermeasures, including drugs and vaccines for smallpox, anthrax, ebola, plague, and other dangerous pathogens. In addition, it will strengthen the National Institutes of Health's capabilities by speeding research and development on medical countermeasures based on promising scientific advances, and give the FDA the ability to make new treatments available in emergency situations (69). Project BioShield was signed into law by President Bush on July 21, 2004. This law provides new tools to improve medical countermeasures that would protect United States citizens against CBRN attack (70).

ANIMAL ACT

On May 30, 2002, the FDA amended the New Drug and Biological Products Regulation so that certain human drugs and biologics intended to reduce or prevent serious or life-threatening conditions may be approved for marketing, based on evidence of their effectiveness from appropriate animal studies and additional supporting data, when human efficacy studies are not ethical or feasible. This rule is part of FDA's effort to help improve the nation's ability to respond to emergencies including terrorist events. The rule was made effective on June 30, 2002. This action was taken in recognition of the need for adequate medical responses to protect or treat individuals exposed to lethal or permanently disabling toxic CBRN substances.

BIOTERRORISM

Toxicologists are more concerned with chemicals and their toxicity, mechanism of action, fate and effects, as well as decontamination. Biological agents also have the same characteristics as hazardous chemicals, but there are significant differences. Primarily, biological agents require an incubation period starting from exposure to the onset of signs and symptoms, and it is important that natural diseases be differentiated from intentional disease outbreaks.

Bioterrorism has been defined as the release of biological agents or toxins that affect humans, animals, or plants, with the intent to harm or intimidate. Biological agents are pathogens that can multiply and cause infections leading to pathogenicity and lethality, while toxins are chemical products of natural origin. Pathogens are considered most threatening when used in biological warfare and terrorism, based on their infectivity, virulence, pathogenicity, lethality, incubation period, contagiousness, and stability. These are designated by the CDC as Category A agents (smallpox, anthrax, plague, botulism, tularemia, and viral hemorrhagic fevers); Category B agents, which include the toxin ricin and other pathogens with lower morbidity and mortality rates than Category A agents, are considered more difficult to disseminate, and are less likely to challenge the Public Health System. Emerging pathogens are defined as Category C agents. The biggest difference between the effects of chemical and biological agents is the time factor. Exposure to chemical agents usually causes immediate effects, while exposure to biological agents has a delayed effect ranging from hours to days, and in some cases weeks. This delayed onset is due to the incubation period between the time of infection and the time for signs and symptoms to develop. Unlike chemical agents, (nerve and blister) biological agents do not penetrate healthy unbroken skin. However, some toxins such as poison ivy and mycotoxins can affect intact skin. Thus to be effective, biological agents must be inhaled, ingested, or penetrate damaged skin. As biological agents do not evaporate, they are dependent on deliberate environmental dissemination (71).

Biological agents are more toxic than chemical agents by weight. Ricin, a biological toxin, is two to three times as toxic as the most toxic nerve agent VX, and botulinum toxin is 5000 times more toxic than VX. Another characteristic of biological agents is that they are invisible to our senses. We cannot see, taste, feel, or smell them, and hence they are difficult to detect (71).

Biological agents have a variety of effects depending on the organism, the dose, and the route of entry. These effects vary from skin and eye irritation to death. These organisms have a natural source, which makes it difficult to distinguish a natural infection from a terrorist-induced infection. Because most of the biological agents are living organisms, they can be destroyed by the environment. Adverse temperature and humidity can affect them. Ultraviolet rays in sunlight will kill most of them within hours, except for anthrax spores that can survive in the environment for years. Because of this, biological attacks are most effective at night or in enclosed areas. Some pathogens (bacteria and viruses) are contagious in that they can be transmitted from person to person and these include pneumonic plague, smallpox, and viral hemorrhagic fevers (e.g., ebola) (71).

The most effective method of delivery of both chemical and biological agents is by the inhalation of an aerosol containing particles in the size range 1 to 5 μm. These can be inhaled into the alveoli during normal respiration. Larger particles can readily fall out of the aerosol or become trapped in the upper airways. Smaller particles are breathed into the alveoli, but can be expired without retention.

Biological warfare agents include bacteria, viruses, and toxins. Bacteria are self-sufficient, independent single-celled living organisms. Spherical bacteria (cocci) range in size from 0.5 to 1.0 μm, and a rod-shaped bacillus can be from 1 to 5 μm long. Bacteria have all of the ingredients necessary to sustain life (cell wall, nucleus containing DNA, and cytoplasm). They reproduce via replication of their own DNA and simple cell division using adenosine triphosphate (ATP) to provide internal energy.

Detection devices for bacteria in the environment are based on the presence of DNA and ATP. For survival in the environment, bacteria require varying conditions of light, moisture, pH, and temperature. The human body provides the ideal conditions for bacteria to thrive. Under special conditions, some bacteria can become dormant by forming spores, which are hard shells, to protect them from adverse environmental conditions until the conditions are suitable for reproduction. Bacteria cause disease either by invading and killing cells or by producing toxic chemicals during their life cycle, or both.

VIRUSES

Viruses are living organisms that require a host organism to survive. They contain genetic materials (RNA or DNA) in a protein shell, but lack the system for metabolism. Thus, they require a living host to provide them basic life support. Viruses reproduce by invading the host cell and by inducing the cell to replicate the virus instead of the host's own DNA, eventually killing the host cell. When the cell dies, the viruses are released from the host to invade nearby cells and the cycle continues. Viruses are smaller than bacteria, ranging in size from 0.02 to 0.2 μm. Most do not contain the double-stranded DNA or ATP, so many biological detection devices cannot detect them. Poxviruses including smallpox consist of double-stranded DNA cross-linked at each end.

TOXINS (BIOTOXINS)

Toxins or biotoxins are toxicants or poisons produced by bacteria, viruses, plants, and animals. They include animal venoms (snake, spider), plant toxins (ricin, poison ivy), and bacterial toxins (botulinum). Because they are actually chemicals, they do not reproduce and are not contagious. Most toxins are proteins, but are considered biological because they are included in the Convention on the Prohibition of the Development, Production and

Stockpiling of Bacteriological (Biological) and Toxin Weapons and on Their Destruction, entered into force March 26, 1975 (72).

Table 2 provides some of the important features of two examples each of bacteria, viruses, and toxins that are considered as potential bioterrorism threats.

The CDC regulates the possession of biological agents and toxins that have the potential to pose a severe threat to public health and safety. CDC's Select Agent Program oversees these activities. The Select Agent Program requires registration of facilities that includes government agencies, universities, research institutions, and commercial entities. The Select Agent regulation Internal Final Rule is documented in 42 CFR 73.0. The CDC Select Agent Program does not release site-specific or identifying information associated with 42 CFR Part 73.0 to the public (71,73–75).

REFERENCES

1. Harris SH. Factories of death: Japanese Biological Warfare 1932–45 and the American Cover-up. New York: Routledge, 1994.
2. Christopher GW, Cieslak TJ, Pavlin JA, Eitzen EM. Biological warfare: a historical perspective. JAMA 1997; 278(5):412–417.
3. Smith RJ. Poisoned wells plague towns all over Kosovo. Aid organizations blame Yugoslav security forces. Washington Post, December 9, 1998, A30.
4. Hickman DC. Seeking asymetric advantage: Is drinking water an Air Force Achilles' heel? Air Command and Staff College, Air University, Maxwell Air Force Base, Alabama. Research Report, 1999, AU/ACSC/084/1999–04.
5. Tucker JB. Historical trends related to bioterrorism: an empirical analysis, 1999. (Available from URL: www.cdc.gov/ncidod/eid/vol5no4/tucker.htm.)
6. Khan AS, Swerdlow DL, Juranek DD. Precautions against biological and chemical terrorism directed at food and water supplies. Pub Hlth Rpt 2001; 116:3–14.
7. Bozeman WP, Dilbero D, Schauben JL. Biologic and chemical weapons of mass destruction. Emerg Med Clin N Am 2002; 20:975–993.
8. Miller JM, Stone, Mitchell C. The Cell. New York: Hyperion, 2002.
9. White SM. Chemical and biological weapons. Implications for anesthesia and intensive care. Br J Anesth 2002; 89(2):306–324.
10. Davis JA. The looming biological warfare storm: misconceptions and probable scenarios. Air Space Power J 2003; Spring. (Available from URL: www.airpower/au.af.mil/airchronicles/apj/apj03/space03/davis.html.)
11. www.terrorism.com/index.shtml
12. www.en.wikipedia.org/wiki/Terrorism
13. Jakucs RM. WMD with toxic industrial chemicals and the Marine Corps response. Marine Corps Gazette, Quantico 2003; 87(4):42–43.
14. www.en.wikipedia.org/wiki/Bioterrorism
15. Ashford DA, Kaiser RM, Bales ME, et al. Planning against biological terrorism: lessons from outbreak investigations. Emerg Infect Dis 2003; 9(5):515–519.
16. www.en.wikipedia.org/wiki/Bioterrorism
17 www.en.wikipedia.org/wiki/Narcoterrorism

18. www.en.wikipedia.org/wiki/Cyber-terrorism
19. Smerconish MA. Flying Blind. Philadelphia, PA: Running Press, 2004.
20. www.army.mil/terrorism/2004–2000/
21. www.en.wikipedia.org/wiki/September_11%2C_2001_attacks
22. www.en.wikipedia.org/wiki/2001_anthrax_attack
23. Coale J. Celebrating 85 years of CB solutions. Edgewood, MD: Edgewood Chemical Biological Center, 2002. (Available from URL: www.ecbc.army.mil/ip/index.htm.)
24. Center for Disease Control and Prevention (CDC). Biological diseases/Agents list, 2002. (Available from URL: www.bt.cdc.gov/agent/agentlist.asp).
25. Burrows WD, Renner SE. Biological warfare agents as threats to potable water. Environ Health Perspec 1999; 107:975–984.
26. Robey J. Bioterror Through Time. Discovery Channel Series, February 21, 2003. (Available from URL: www.dsc.discovery.com/anthology/spotlight/bioterror/history/history.html 3p.)
27. Tschanz DW. "Eye of Newt and Toe of Frog": Biotoxins in Warfare. Strategy Page.com, October 20, 2003. (Available from URL: www.216.239.37.104/search?q=cache:Tv-SBKjTGFIJ:www.strategypage.com/articles/biotoxin_files/BI 3p.)
28. Noji EK. Bioterrorism: A 'New' Global Environmental Health Threat. Global Change & Human Health 2001; 2(1):46–53.
29. United Kingdom Ministry of Defense. Defending Against the Threat of Biological and Chemical Weapons. Outline History of Biological & Chemical Weapons, 1999. (Available from URL: www.mod.uk/issues/cbw/history.htm 5 p.)
30. Smart JK. History of Chemical and Biological Warfare: An American Perspective. In: Textbook of Military Medicine: Medical Aspects of Chemical and Biological Warfare. Sidell FR, Takafugi ET and Franz DR, eds. Office of the Surgeon General, Department of the Army, Washington, DC, Chapter 2, pp 9–86, 1997. (Available from URL: www.usuhs.mil/cbw/history.htm.)
31. Dire DJ. Biological Warfare. eMedicine, Instant Access to the Minds of Medicine, 2003. (Available from URL: www.emedicine.com/aaem/byname/biological-warfare.htm 2p.)
32. The President's Commission on Critical Infrastructure Protection (PCCIP). Critical Foundations: Protecting America's Infrastructures. Critical Infrastructure Assurance Office (CIAO), 1997. (Available from URL: www.ciao.gov/resource/pccip/report_index.htm 190 p.)
33. Frist W. When Every Moment Counts: What You Need to Know About Bioterrorism from the Senate's Only Doctor. Rowman & Littlefield, Inc., New York, 2002:182.
34. Corey R. Short Course Examines Safety of Water Supply. TimesDaily.com, Alabama, 2002. (Available at URL: www.timesdaily.com/apps/pbcs.dll/article?Date=20020813&Category=NEWS&ArNo=208130324&SectionCat=&Template=printart 1 p.)
35. Oehler GC. The Threat of Nuclear Diversion. In: Speeches and Testimony, Statement for the Record by Dr. Gordon C. Oehler, Director, Nonproliferation Center to the Senate Armed Services Committee, "The Continuing Threat from

Weapons of Mass Destruction" 27 March 1996. Central Intelligence Agency. (Available from URL: www.odci.gov/cia/public_affairs/speeches/1996/go_toc_032796.html pp. 3–4.)

36. Bogen DC, Krey PW, Volchok HL, Feldstein J, Calderon G, Halverson J, Robertson DM. Threat to the New York city water supply—plutonium. Sci Total Environ 1988; 70:101–118.

37. Purnick J. Plutonium Found in City Water; Koch Says Supply System is Safe. New York Times, Late City Final Edition. 1985; July 27, Section 1, pp 1.

38. Brennan, Phil (2001). French Terrorist Chief Warns of New Attacks, Possibly Cyanide. Reprinted from NewsMax.com, Tuesday, November 27, 2001. (Available from URL: www.newsmax.com/cgi-bin/printer_friendly.pl?page=www. newsmax.com/archives/articles/2001/11/26/173332.shtml 2 p.)

39. Isenberg D. (2002). Securing U.S. Water Supplies. July 19, 2002. Center for Defense Information, Terrorism Project. (Available from URL: www. cdi.org/terrorism/water.cfm 3 p.)

40. Anderson C. FBI says terrorists may target U.S. food, water. Philadelphia Inquirer 2003; 5:A13.

41. Salem H, Gargan II TP, Burrows WD, Whalley C, Wick CH. Chemical and Biological Warfare Threats to Military Water Supplies, 2004. ECBC-TR-366.

42. Deininger RA. The threat of chemical and biological agents to public water supply systems. Water pipeline database. TSWG Task T-1211, SAIC Contract N39998–00, 2000.

43. Cooke, R. Cybersecurity now! Homeland Defense J 2002; 1(2): 18–23.

44. Cai Z. Cyanide. In: Wexler P, ed. Encyclopedia of Toxicology, Vol. 1. New York: Academic Press, 1998; 387–389.

45. Kowalski W. The Tylenol murders, 1997. (Available from URL: www. personal.psu.edu/users/w/x/wxk116/tylenol.)

46. Fearful. Was the person behind the Tylenol poisonings in the 1980's ever caught? 2003. (Available from URL: www.ask.yahoo.com/ask/20030/30.html.)

47. Mylroie L. The World Trade Center bomb: Who is Ramzi Yousef? And why it matters. The National Interest, Winter, 1995/1996. (Available from URL: www.fas.org/irp/world/iraq/956-tni.htm.)

48. Hammer, J. Code blue in Jerusalem. Newsweek, July 1, 2000, 24–29.

49. Biowar Timeline. 2001. Red Lies. (Available from URL: www.hindustantimes. com/nonfram/191001/detFOR07.asp and www.tv.cbc.ca/national/pgminfo/redlies/time.html.)

50. Arizona Star. A list of anthrax deaths, infections, and locations, 2001. (Available from URL: www.azstarnet.com/attack/anthraxsofar.html.)

51. Boston.com. Anthrax deaths, infections, 2001. (Available from URL: www. boston.com/news/packages/anthrax/cases.htm.)

52. Kaplan DE, Marshall A. The Cult at the End of the World. New York: Crown Publishers, Inc, 1996.

53. Gurr N, Cole B. The New Face of Terrorism. New York: J.B. Tauris Publishers, 2000.

54. www.marsearchconnection.com/okc.html

55. Ohbu S, Yamashina A. Sarin poisoning on Tokyo subway. South Med J 1997; 90(6):587–593.

56. Martin P. Was the U.S. Government alerted to September 11 attack? MSNBC, 2002. (Available from URL: www.wsw.org/articles/2002/jan2002/sept-j16.shtml.)

57. Office of Techology Assessment. Technology against terrorism. Structuring security. OTA-Isc-511, 1995, NTIS order # 152529.

58. Rothstein L. Anthrax hoaxes: hot new hobby? Bull Atomic Sci 1999; 55(4):7–13. (Available from URL: www.thebulletin.org/issues/1999/ja99bulletins.html).

59. Kristoff ND. The anthrax file. New York Times Opinion. July 12, 2002. (Available from URL: www.truthout.org/docs_02/07.13E.kris.anthrax.htm.)

60. www.thebulletin.org/issues/1999/ja99bulletin.html

61. www.truthout.org/docs.02/07.13E.kris.anthrax.htm

62. BBC. US expands "axis of evil." BBC News, May 6, 2002b. (Available from URL: www.news.bbc.co.uk/1/hi/world/americas/1971852.stm.)

63. NPS. Terrorist group profiles. Index of groups. Dudley Knox Library, Naval Postgraduate School, 2003. (Available from URL: www.library.nps.navy.mil/home/tgp/tgpndx.htm.)

64. Carnell B. Opening statement by the FBI's James F. Jarboe on ecoterrorism, Feb. 13, 2002. (Available from URL: www.animalrights.net/articles/2002/000058.html.)

65. www.whitehouse.gov/homeland/book/index.html

66. www.clerk.house.gov/evs/2001/roll398.xml

67. www.senate.gov/legislative/LIS/roll_call_lists/roll_call_vote_cfm.cfm?   con gress=107&session=1&vote=00313

68. www.en.wikipedia.org/wiki/USA_PATRIOT_Act

69. www.whitehouse.gov/news/release/2003/02/20030203.html

70. www.whitehouse.gov/bioshield

71. Kortepeter, M. USAMRIID's Medical Management of Biological Casualties Handbook, 2001.

72. www.state.gov/t/ac/trt/4718.htm

73. www.cdc.gov/od/sap

74. Weintraub, P. Bioterrorism: How to Survive the 25 Most Dangerous Biological Weapons. New York: Citadel Press, 2002.

75. Acquista A. The Survival Guide. In: What to Do in a Biological, Chemical, or Nuclear Emergency. New York: Random House, 2003.

3

The Food Quality Protection Act and the Safe Drinking Water Act[a]

William L. Jordan

U.S. Environmental Protection Agency, Office of Pollution Prevention and Toxic Substances, Washington, D.C., U.S.A.

Gary E. Timm

U.S. Environmental Protection Agency, Office of Science Coordination and Policy, Washington, D.C., U.S.A.

INTRODUCTION

Pesticides are widely used in the United States, particularly in agriculture. Over the last four decades, public controversy has swirled around the government's regulatory programs, usually in connection with specific pesticides such as dichloro diphenyl trichloroethane (DDT) in the 1960s; Alar®, the plant growth regulator used on apples featured on the CBS newsmagazine "60 Minutes" in the 1980s; and the genetically modified variety of corn, StarLink®, that was illegally introduced into the human food supply. As a consequence, government agencies responsible for regulating pesticides have had to deal with a broad array of cutting-edge scientific issues. These issues range from setting the government's policies on cancer risk assessment to pioneering work in the field of ecological risk assessment, to defining for the country and for the world how to evaluate the safety of the products of modern biotechnology. This chapter focuses

---

[a]The views expressed in this chapter are the views of the authors and do not necessarily represent the policies of the U.S. Environmental Protection Agency.

on the latest addition to this long history of scientific accomplishments—the validation of new assays for evaluating the effects of substances on the endocrine system and the U.S. Environmental Protection Agency's (EPA) pioneering efforts to improve chemical risk assessment methodology.

In 1996, Congress passed two extraordinary pieces of legislation affecting the regulation of pesticides and other chemicals. These laws, the Food Quality Protection Act (FQPA) and the Safe Drinking Water Act Amendments of 1996 (SDWA Amendments), forced EPA to integrate a new level of scientific advances into its regulatory programs for pesticides and drinking water contaminants, and, because of the forward-looking provisions of these laws, to push the state of the science in the emerging fields of endocrine disruption and risk assessment. The story of FQPA reveals how the scientific world shaped legislation and how legislation, in turn, is stimulating major scientific advances.

FQPA is the result principally of three different causes: the controversy over implementation of the anticancer Delaney clause in the Federal Food, Drug, and Cosmetic Act (FFDCA), which culminated in federal court decisions that effectively required EPA to prohibit numerous uses of carcinogenic pesticides; the ground-breaking report from the National Academy of Sciences (NAS), "Pesticides in the Diets of Infants and Children," calling for significant changes to EPA's risk assessment methodologies; and the publication of Dr. Theo Colburn's book, *Our Stolen Future*, that brought popular attention to the potential of pesticides and other environmental contaminants to cause adverse effects on the functioning of endocrine systems in humans and wildlife.[b] Many of the landmark provisions of FQPA found their way into the SDWA Amendments.

THE "DELANEY CLAUSE" IN THE FFDCA

FQPA amendments passed in 1996 were, in large measure, a legislative response to a scientific controversy that arose several generations earlier. In 1954, Congress added a provision to the FFDCA that required the Food and Drug Administration (FDA) to establish "tolerances" or maximum permissible levels for the residues of pesticides in "raw agricultural commodities."[c]

---

[b] A law review article, "All the Stars in the Heavens Were in the Right Places: the Passage of the Food Quality Protection Act of 1996," by James Smart, traces, in much greater detail, the long, convoluted history of legislative efforts to rewrite the pesticide law. 17 Stanford Environmental Law Journal 273 (1998).

[c] The government also regulates pesticides under the Federal Insecticide, Fungicide, and Rodenticide Act (FIFRA). Under FIFRA, the government regulates the sale and use of pesticide products. Government policy coordinates decisions under the FFDCA and FIFRA. Sale and use of a pesticide product will be approved only if any pesticide residues expected to be present in food are authorized under the FFDCA. Conversely, if residues of a pesticide may not be approved in food, the use of the pesticide causing such residues will not be allowed. In 1996, EPA articulated this policy in a Federal Register notice, although EPA had followed it for many years. 61 Fed. Reg. 2378.

The FFDCA declared a food adulterated if it contained either a pesticide for which there was no tolerance specified or quantities of a pesticide that exceeded its specified tolerance. For setting a tolerance, FDA had to determine whether the residues were "safe," by weighing the risks of consuming the product against the benefits of using the pesticide to grow the commodity. In 1958, Congress added another provision subjecting "food additives," substances that were intentionally added to food, to a similar regulatory scheme. The presence of an unapproved food additive rendered food adulterated and liable for penalties and seizure by FDA.

Significantly, the safety standard in the food additive provision was more stringent than the pesticide provision's risk–benefit balancing test. The food additive provision simply contained language requiring a safety finding to authorize the presence of an additive in food. The legislative history explained that "safe" meant "a reasonable certainty of no harm." This Congressional explanation precluded the consideration of the benefits of using the food additive. In addition, the statutory provision prohibited the authorization of an additive in food if the substance had been found to "induce cancer" in humans or animals. This provision, called the Delaney clause after the Congressman who proposed the provision following his wife's death from cancer, reflected the scientific understanding of the day, which did not distinguish among animal carcinogens on the basis of either their potency or their human relevance. Thus, the food additive provision of the FFDCA sets a strict ban on carcinogenic food additives, regardless of the degree of risk they might pose.

The food additive and pesticide provisions of the FFDCA did not fit neatly together, despite Congressional efforts to coordinate the two sections. In 1958, Congress recognized that many processed foods would still contain at least some of the pesticide residue initially present on the raw agricultural commodities from which they were made. To reconcile the food additive provision with the pesticide provision, Congress expressly excluded the presence of pesticide residue on a processed food from the broadly phrased definition of food additive, so long as it fell below the tolerance for the pesticide on the corresponding raw agricultural commodity. This exclusion effectively meant that, most of the time, only the pesticide provision of the FFDCA governed decisions about the safety of pesticide residues in food. But, certain pesticide uses were still subject to the food additive provision because some processed foods will contain higher levels of a pesticide than allowed by the raw agricultural commodity tolerance due to the concentration of residues during cooking, dehydration, or other processing of the commodity. For example, when processors make tomato paste from fresh tomatoes, they cook the tomatoes at high temperatures removing a large percentage of the water. If the pesticide residue is stable when heated and binds to the tomato solids, it usually will concentrate in the paste. Many other processed food and feed forms—oils, dried vegetables, for instance—will also bear higher concentrations of many pesticides than the raw commodities from which they were made. These concentrated pesticide residues in processed food were subject to the food additive provision and the Delaney clause. This scheme led to the striking paradox—the government could declare a very low level of a carcinogenic pesticide residue in a fresh tomato to be "safe" under the FFDCA's pesticide provision, but exactly the same quantity of the pesticide in tomato paste (at a higher concentration) could run afoul of the Delaney clause, causing the residue to be "unsafe."

After its formation in 1970, responsibility for administration of the pesticide provision of the FFDCA passed to the EPA. To make these legislative provisions work smoothly, EPA adopted its "coordination policy" under which it would not establish a section 408 tolerance for a pesticide that needed, but could not receive, a section 409 food additive regulation. Thus the Delaney clause and EPA's "coordination policy," together, under the FFDCA, prohibited the establishment of food additive approvals and associated tolerances for a carcinogenic pesticide if the residues allowed on a raw agricultural commodity could concentrate in any form of processed food derived from that commodity. During the 1970s and 1980s, EPA applied this approach prospectively—that is, to the review and approval of both new pesticides and new uses of previously approved pesticides. During this same period, pesticide companies, charged with updating the toxicity databases on their products, performed new studies, some of which identified existing pesticides as animal carcinogens. When EPA obtained reports from a new research identifying a pesticide as a carcinogen, the Agency evaluated the potential human risks using more advanced risk assessment methods than those available in the 1950s. In many cases, EPA decided that these pesticides did not pose a risk to public health, either because the animal carcinogens were unlikely to affect humans, or the potential human risk was inconsequential because the carcinogen was not very potent or residues were negligible. In such situations, EPA did not revoke food additive approvals that had previously been established, even though the Delaney clause arguably compelled that course.

In response to mounting pressure for action, from environmental and public health advocacy groups, in 1985 EPA asked the NAS to study the likely impacts of implementing the Delaney clause on public health. In its 1987 report, "Regulating Pesticides in Food—the Delaney Paradox" (Delaney Paradox report), the NAS concluded that the application of the Delaney clause had decreased overall public health protection, and likely would continue to do so. This outcome occurred because the Delaney clause led EPA to approve the use of noncarcinogenic pesticides to address agricultural pest problems that could have been effectively mitigated through the use of alternative pesticides that were less risky, but carcinogenic. The NAS concluded that a single "negligible risk" safety standard, applied evenhandedly to all types of pesticide residues in all types of food, would achieve the greatest level of public health protection while concomitantly maintaining most of the socially and economically beneficial uses of pesticides.

Notwithstanding the NAS's Delaney Paradox report, environmental groups continued to press EPA to apply the Delaney clause to old, carcinogenic pesticides. In response, EPA began to explore a variety of ways to reduce the impact of the Delaney clause. A few EPA policies had only limited applicability, for example, approving a pesticide for use on a raw commodity with a restriction that the commodity could not be sent for processing. But in 1988, EPA advanced a major policy initiative to implement the NAS recommendation: EPA asserted a "de minimis" exception to the Delaney clause. In simple words, this initiative argued that the FFDCA should and could be read in a flexible, commonsense way, rather than strictly, especially when a literal application of the law would produce absurd results. Thus, under this legal doctrine, EPA believed that it could lawfully authorize residues of a carcinogenic pesticide in a processed food if the carcinogenic risks from that residue were negligible or "de minimis." After all, as the NAS had suggested, a prohibition on the carcinogen might actually increase public health risks.

Environmental and public health advocacy groups successfully challenged EPA's "de minimis" policy in federal court in a case involving existing food additive regulations for carcinogenic pesticides. In 1992, the U.S. Court of Appeals for the Ninth Circuit unanimously ruled in the case, *Les v. Reilly*, that Congress intended the Delaney clause to be applied exactly as written, and disallowed EPA's policy (968 F.2d 985). Following the *Les* decision, EPA began to implement the Delaney clause by proposing to revoke the food additive approvals for carcinogenic pesticides, although EPA's progress was still too slow to satisfy environmental groups.

In 1995, EPA agreed to settle another lawsuit filed earlier, *NRDC v. Browner*, brought by the same coalition of environmental groups who sought to compel the agency to apply the Delaney clause and the coordination policy to old, carcinogenic pesticides that concentrated in processed food but did not come under the necessary food additive regulations. Under the settlement, EPA agreed to an aggressive schedule for reviewing data on pesticide concentration, and if necessary, initiating proceedings under the FFDCA to revoke any approvals not comporting with the coordination policy. Of course, EPA would also need to prohibit future use of pesticides for which it had revoked tolerances and food additive approvals. By EPA's estimate, there were over 30 pesticides identified as inducing cancer with about a hundred associated tolerances and food additive approvals. Needless to say, the companies manufacturing such pesticides and the farmers who used them were quite alarmed at the prospect of the impending EPA actions.

THE NAS REPORT ON "PESTICIDES IN THE DIETS OF INFANTS AND CHILDREN"

The second major factor leading to the passage of the FQPA was the issuance of a ground-breaking report entitled, "Pesticides in Diets of Infants and Children," by the NAS in the late fall of 1993. Commissioned by the Congress in 1988, in large part at the behest of environmental advocacy groups, this report focused on the scientific and policy issues involved with regulating the exposure of infants and children, to pesticides in their diets. Led by Philip Landrigan, a distinguished panel of scientists and public health experts reviewed the emerging body of scientific evidence concerning the potential effects of pesticides on children. In summary, the report concluded that EPA's then-current approach to assessing risks to children and infants had failed to keep pace with scientific developments in a number of areas, and therefore EPA's regulatory programs might not provide adequate protections for public health. NAS called for major changes at EPA.

NAS found many differences between children and adults that are significant for risk assessment. In terms of age-related variations in susceptibility and toxicity, the NAS concluded that there was ample evidence of quantitative and occasional qualitative differences in the toxicity of pesticides to children and adults. In other words, infants and children are not simply "little adults." Although sometimes children can be less sensitive or equally sensitive to the harmful effects of a toxicant, at times they are more sensitive than adults. NAS found that the quantitative differences in toxicity were usually less than approximately 10-fold. NAS also noted significant differences in the exposure of infants and children compared to adults. The NAS's analysis of dietary consumption survey data showed that throughout their early years, children ate more food per unit of body weight as well as a different mix of foods when compared to adults. Therefore most children in preschool or elementary school would be taking in more pesticide residue relative to their body weight through their daily diets than the typical grownup. Additionally, the NAS pointed out that there is considerable variation in the levels of pesticide residue in raw agricultural commodities, as well as a wide range in the amount of different commodities consumed by individuals. These sources of variation meant that, on a given day, there could be an extremely large range of dietary exposures to a particular pesticide for different members of the public.

These conclusions led the NAS to recommend significant changes to the approach EPA had used to assess dietary risks of pesticides. The changes included are given below:

> *Exposure estimates*: Assessments should reflect the unique characteristics of the diets of infants and children and should account for all nondietary sources of exposure to a pesticide, such as through drinking water or as a consequence of pesticide use in and around homes, schools, playgrounds, and other public areas.

Toxicity studies: Assessments should include sufficient testing to allow evaluation of the potential differences in susceptibility and sensitivity between adults and the young, including improved reproductive toxicity studies and studies to measure the relative sensitivity of adults and the young.

Uncertainty factors: Determinations of a safe level of exposure to a pesticide should include an additional uncertainty factor of up to 10-fold whenever there is an evidence of postnatal developmental toxicity or when data on the sensitivity of children are incomplete. This uncertainty factor would account for the possibility that infants and children may be more adversely affected by exposure to a toxicant than an adult. (Unfortunately, the NAS incorrectly described this as EPA's then-current practice.)

Food consumption data: Due to the significant limitations of the food consumption surveys available in 1993, EPA should obtain and use data that better characterize the unique patterns of dietary intake of infants and children, and these data should be updated periodically.

Pesticide residue data: Due to limitations in the quantity and quality of data on the distribution of pesticide residues in food, the government should conduct more extensive monitoring of foods, particularly those important to the diets of children, and collect and analyze samples closer to the point of consumption.

Risk assessment methods: These should include, in addition to exposure from food, all other pathways by which the public may potentially be exposed, particularly through drinking water; should consider exposures to multiple chemicals causing a common toxic effect; and should develop "probabilistic" risk assessments that account for variability in exposures and risks across significant subgroups, particularly age-based subgroups.

EPA responded to the NAS report by making a promise to move in the recommended directions. Several initiatives had already been planned, and the government carried them forward. For example, the U.S. Department of Agriculture (USDA) had planned and began, in 1994, to execute a comprehensive, three-year, national survey of food consumption, the Continuing Survey of Food Intake by Individuals (CSFII). The new CSFII survey employed numerous quality control steps that addressed many of the limitations identified in the NAS report. In addition, in 1998, USDA performed a supplemental survey that focused on children up to the age of eight years, thereby significantly expanding the size of the databases available to EPA for estimating dietary pesticide exposure to children.

Although initially mandated before publication of the 1993 NAS report, USDA's Pesticide Data Program (PDP) quickly came to serve as the authoritative source of information on the levels of pesticide residues in food. Operated as a state–federal partnership, the PDP collects samples of targeted commodities throughout the year and across the country so that the values provide a representative picture of the frequency and levels of pesticides in food. Following the suggestion of the NAS, PDP focuses particularly on the most important sources of dietary exposure in children, primarily fruits and vegetables. The samples are collected at wholesale distribution points, just before they go on retail sale to the ultimate consumers.

While the USDA PDP database and the CSFII survey data eventually came to play a critical role in EPA's risk assessment methodologies, EPA did not move particularly swiftly with regard to many of the other NAS recommendations. Due to lack of resources, lack of clear pathways for implementing the Academy's ideas, and few strong forces compelling a change, EPA let most of the other recommendations languish.

There was one notable exception. EPA developed a new set of guidelines for conducting reproductive toxicity testing in mammals. This new guideline provided nonbinding recommendations with regard to pesticides and industrial chemicals to organizations performing reproduction studies. Among the improvements were an expanded list of measurements to be recorded during the in-life phase of the test and more extensive postmortem examination of the test animals.

THE CONCERN FOR ENDOCRINE DISRUPTORS

During the 1970s scientists became aware of the potential for certain environmental contaminants and pharmaceuticals to interact with the estrogen receptor (ER) and thereby affect human health as well as wildlife. Concerns regarding endocrine disruptors are primarily due to (i) adverse effects observed in certain wildlife, fish, and ecosystems, (ii) increases in some endocrine-related human diseases and developmental abnormalities, and (iii) observations that chemicals can interact with the endocrine system of experimental animals to interfere with normal reproduction, development, and growth.

The Wingspread Conference, held in 1991, under the topic "Chemically Induced Alterations in Sexual and Functional Development: The Wildlife/Human Connection," served as a catalyst to research; it brought together people of various disciplines who shared their perspectives and developed a broader, more coherent picture of the problem of chemical effects on the developing embryo (1).

> The Wingspread I consensus position had taken a significant leap forward, both in the breadth of its hypothesis connecting wildlife abnormalities to human risks and in the confidence of the assertion that this was a real effect. Not only did the consensus statement [issued by conference participants] bring cohesion to workshop participants, but it provided a highly effective tool for raising the issues

before a nonscientific audience. Less than three months after the Wingspread I meeting, the consensus position was reported to a Senate committee investigating reproductive hazards. The consensus statement was subsequently cited in many journalistic accounts and brandished by activists as a new call to arms against environmental contaminants (2).

Two congressional hearings advanced the issue toward legislative consideration. The first a Senate hearing in 1991, focused generally on reproductive hazards. The second, held in the House in 1993, focused on the Health Effects of Estrogenic Pesticides and featured many of the scientists who had participated in the Wingspread Conference. In subsequent years, there were numerous bills dealing with the testing and research of endocrine disruptors, but none were enacted until the summer of 1996.

FQPA OF 1996

By the spring and summer of 1996, an unexpected confluence of forces had come to bear on the various stakeholder communities interested in pesticide regulation that would make the passage of new legislation possible. In the past, either the pesticide industry and their allies in the agricultural community or the environmental and public health advocacy groups had the ability to block legislation which they considered unacceptable. Now, however, all of the stakeholders felt dissatisfied with the status quo and felt the need for something that only Congress could provide. The pesticide industry and pesticide users were alarmed at the prospect of full implementation of the Delaney clause as required by the Ninth Circuit's decision in *Les v. Reilly*, and promised in the settlement of *NRDC v. Browner*. Not only would the Delaney clause compel the removal of large numbers of pesticide uses from the marketplace, but its continued existence would also hang like a dark cloud over all future pesticide uses. The public interest community, on the other hand, was very frustrated with the EPA over its lack of progress in upgrading its regulatory science, and in their eyes, continuing to make safety decisions that ignored obvious and serious risks.

The stakeholders' differing perspectives laid the basis for a deal: Congress would rewrite the FFDCA to eliminate application of the Delaney clause to pesticides and would replace it with language that compelled EPA to move ahead with the recommendations in the 1993 NAS report. In addition, the environmental advocacy groups wanted an aggressive program to look at the potential endocrine disrupting effects of pesticides. EPA, FDA, and USDA all agreed that the Administration should support new legislation with these goals.

Crafting legislation from scratch that would achieve all these ends would have been impossible in the space of a few months, but fortunately a strong foundation already existed. Meeting under the sponsorship of the Keystone Center from 1990 to 1993, a group of individuals from organizations across the spectrum of pesticide interests discussed how to move ahead constructively in regulating pesticides. In September 1993, this group issued its consensus conclusions in a report titled, "The Keystone National Policy Dialogue on Food Safety and Pesticides." This report recommended a range of legislative and administrative changes, including replacing the Delaney clause with a safety standard that considered only risks, but not benefits, for pesticide residues in food. Building on this consensus, staff from key Congressional committees and from EPA, USDA, and FDA had drafted a bill that reflected these basic directions.

With Presidential and Congressional elections looming, everyone recognized the need to move quickly. The key Congressional committees summoned all of the stakeholders and asked for a consensus bill. Facing a short deadline, the stakeholders accepted the staff's working draft, essentially in its entirety, and it became possible to consider enacting sweeping amendments to the FFDCA. FQPA of 1996 had undergone three committee mark-ups and passed both houses of Congress, astonishingly, without a single dissenting vote and in only three weeks. President Clinton signed the law on August 3, 1996, amid a euphoric crowd of senior officials from the pesticide industry, farm organizations, and environmental and public health groups, as well as members of Congress and the Administration.

FQPA was revolutionary. Its amendments to the FFDCA established a single set of substantive standards and procedures applicable to all tolerance-setting decisions. FQPA replaced the bifurcated standards (risk–benefit balancing for pesticides in raw agricultural commodities and a risk-only standard for pesticide residues that concentrated in processed foods) with a single, risk-only safety standard for all types of foods. Its truly extraordinary provisions, though, were the ones elaborating on the meaning of the new safety standard:

> *Aggregate risk*: FQPA defined safety as "a reasonable certainty that no harm will result from aggregate exposure to the pesticide chemical residue, including all anticipated dietary exposures and all other exposures for which there is reliable information." This language, in effect, compelled EPA for the first time to do what the NAS had recommended several years earlier—to add up all the exposures a person could receive from pesticides in food, in drinking water, and in and around places they lived, worked, or played. (Occupational exposure from mixing, loading, or applying pesticides was not included in aggregate risk.)
>
> *Cumulative risk*: FQPA directed EPA to consider, among other factors, "available information concerning the cumulative effects of such residues and other substances that have a common mechanism of toxicity." This provision, too, responded directly to an NAS

recommendation in 1993 that EPA take into account the combined impact of exposure to multiple pesticides that have similar effects on the human system.

Sensitive subpopulations: FQPA directed EPA to consider, among other factors, "available information concerning the variability of the sensitivities of major identifiable subgroups of consumers."[d]

Children's Safety Factor: FQPA required that, when evaluating the safety of pesticide residues in food, "an additional 10-fold margin of safety for the pesticide chemical residue and other sources of exposure shall be applied for infants and children to take into account potential pre- and postnatal toxicity and completeness of the data with respect to exposure and toxicity in infants and children." The FQPA, however, did allow EPA to employ a different safety factor, but "only if, on the basis of reliable data" the resulting margin of safety would be adequate for infants and children.

Endocrine Disruptor Screening Program (EDSP): FQPA required EPA to develop and, within three years, begin to implement a program to screen all pesticide chemicals, both active and inert ingredients in pesticide formulations, for any "effect in humans that is similar to an effect produced by a naturally occurring estrogen and such other endocrine effect" as EPA may designate.[e]

Two other features of FQPA were also extremely significant. First, it took effect immediately; there was no delayed effective date at which EPA could begin preparing itself and the regulated community to meet the new standards. Second, FQPA applied not only to decisions made on proposed new tolerances but also, through a program called "tolerance reassessment," to all tolerances in existence when FQPA became a law. FQPA required EPA to re-evaluate over 9600 tolerances in force in 1996 on a 10-year schedule. To assure steady progress in meeting the full reassessment goal, EPA had to reassess 33% within the first three years, a total of 66% by August 2002, and all 100% by August 2006. The tolerance reassessment requirement, thus, was designed to ensure that all pesticide tolerances would meet the new safety standard within 10 years. In combination, these different provisions launched a whirlwind of change in the regulatory science used by EPA to regulate pesticides.

---

[d] The SDWA Amendments added a similar provision: By 2000 and periodically thereafter, EPA was to conduct studies to identify subpopulations at greater risk (e.g., infants, children, pregnant women) than the general public of adverse health effects from exposure to contaminants in drinking water and to report the results of those studies to Congress.

[e] Also in the fall of 1996, Congress enacted the Safe Drinking Water Act Amendments of 1996. These amendments included a provision giving EPA discretion to require endocrine disruptor testing of substances that may contaminate sources of drinking water. This provision, in effect, expanded the scope of chemicals for which EPA may potentially require screening studies.

As mentioned in the footnotes, several of the landmark provisions from FQPA found their way into the SDWA Amendments. The provision on endocrine disruptors in FQPA became a section of the SDWA that authorized testing of substances that may be found in sources of drinking water to which a substantial population may be exposed. Similarly, the emerging recognition that people may differ markedly in their sensitivities to adverse effects of environmental pollutants shows up both in FQPA as a factor EPA must consider in setting tolerances and as a mandate in SDWA for EPA to conduct research on age and other factors that may lead to differing susceptibility to chemicals' effects. Of course, the SDWA Amendments made many other significant changes and additions to the SDWA. These included new prevention approaches, improved information to consumers, changes designed to strengthen the regulatory program, and new funding for states and local programs.

IMPLEMENTING THE FQPA REQUIREMENTS FOR AGGREGATE RISK, CUMULATIVE EXPOSURE, AND THE CHILDREN'S SAFETY FACTOR

Shortly after passage of the FQPA, EPA established an advisory committee, the Food Safety Advisory Committee (FSAC), to provide advice on the implementation of the new law. Among other suggestions, a number of the FSAC members suggested that EPA should engage in a notice-and-comment process to announce the Agency's approach to several of the key scientific provisions. Working closely with its independent external advisory committee of scientific peer reviewers, the FIFRA Scientific Advisory Panel (SAP), EPA crafted detailed science policy documents that described how the agency addressed such issues as aggregate risk assessment, cumulative risk assessment, and the FQPA Children's Safety Factor. In addition, EPA began applying these policies to specific chemicals as it reviewed new tolerances and reassessed existing tolerances. EPA referred a number of these chemical-specific assessments to the SAP for further examination. The following paragraphs describe some of the scientific advances, primarily, in the area of improved risk assessment methodologies, resulting from these efforts.

Probabilistic risk assessment of single-day exposures to pesticides in the diet: Because people weigh different amounts, eat different foods, eat different amounts of food, and because individual servings of the same food may contain different levels of pesticide residue, potential exposure to pesticides in the diet can vary quite significantly across the population. Working initially with a private corporation, Novigen, Inc., EPA began to develop probabilistic risk assessments for acute dietary exposure that could take all of these sources of variability into account. EPA used data from USDA's CSFII and PDP to generate a distribution of daily pesticide intakes. The method, made possible only through significant advances in the quality of available databases and computing power, involved calculating a daily exposure for each of the approximately 15,000 individual respondents surveyed in the CSFII by multiplying the amount of each commodity the respondent consumed by a randomly selected residue value measured for that commodity by PDP. By repeating this process approximately a thousand times for each individual, and then combining all calculated exposures, the method generated a very stable distribution that EPA used to represent exposure in the general public and in significant subgroups.

In addition to calculating distributions of exposure, EPA needed to decide how to interpret these probabilistic risk assessments in light of the new FQPA safety standard. After gaining some experience with performing probabilistic exposure analyses, the Agency decided to refer the 99.9th percentile of the estimated exposures as the level corresponding, in most cases, to "a reasonable certainty of no harm." Under this policy, the estimated exposure for at least 99.9% of the population would be less than the value associated with this percentile. If that level of exposure were considered acceptable, EPA concluded that it could declare that the residue would be safe.

Aggregate exposure estimates: FQPA required EPA, in making its safety determinations for a tolerance, to consider all potential, nonoccupational sources of exposure to a pesticide. In general, these included exposure through food, through drinking water, and from residential use of pesticides. (EPA used the term "residential use" to include pesticide use in and around homes, schools, offices, parks and playgrounds, and other public places where a person might encounter a nonagricultural pesticide.) Prior to FQPA, EPA had generally prepared separate estimates for each of these pathways and only combined them occasionally. Because the estimate for each pathway was conservative, that is, designed not to underestimate potential exposure, EPA thought that combining the separate estimates would result in a significant overestimation of the potential exposure. It was, in EPA's view, quite unlikely that any person would encounter the highest level of residues both in his daily diet and in his drinking water on the same day, while simultaneously experiencing the upper end of potential residential exposure.

FQPA forced the development of more sophisticated modeling approaches. Within eight years after FQPA became a law, three new, probabilistic, multiple pathway exposures had been created. The first model, an outgrowth of Novigen's probabilistic food exposure model, was called CALENDEX™. The second model, LifeLine™, was developed through a grant from EPA. A consortium sponsored by Crop Life America, the trade association for the pesticide industry, developed the third model, referred to as CARES™. Comparisons of the three models reveal that on using the same input data they produce remarkably similar results, all substantially lower than the values that would have been derived by adding the separate high end estimates for each separate pathway.

Cumulative exposure estimates: EPA has identified several groups of chemicals that share a common mechanism of toxicity. The largest group is the organophosphate pesticides (OPs) which all inhibit acetylcholinesterase, an enzyme essential for the proper functioning of the nervous system. EPA identified 31 different OPs to assess in the Cumulative Assessment Group (CAG). The OP risk assessment was in some ways quite straightforward; the two fundamental tasks were (i) to determine how much concurrent exposure to the different OP chemicals did people receive and (ii) to determine how to account for the differing levels of risk posed by different compounds when concurrent exposure occurs. The preparation of the cumulative risk assessment for OPs, nonetheless, involved several years of effort and resulted in the most complex and sophisticated chemical risk assessment prepared to date by EPA's pesticide program.

EPA tackled the second problem first. Because the toxicity of the different members of the CAG varied by several orders of magnitude (more than 1000-fold), EPA needed to develop a basis for combining the potential exposures to different OP compounds. EPA did so by estimating relative potency factors (RPFs) for each chemical in the CAG. The calculation of the RPF was based primarily on estimates for each chemical of its ED_{10} (the dose at which 10% of the exposed organisms would be expected to display a statistically significant level of cholinesterase inhibition) following multiple days of exposure. This task was complicated by the need to work with multiple studies for a single chemical, which often differed in dose levels, duration, and other critical factors.

Once EPA calculated the RPFs, it then needed to estimate how much exposure people received to different OPs. This was a daunting challenge—because like the individual chemical and aggregate risk assessments, it was essential both to capture concurrent exposures and to avoid combining exposures that were unlikely ever to co-occur. To generate realistic exposure scenarios, EPA developed 11 regional risk assessments that reflected geographic and temporal variability in the use of different OPs.

For each regional estimate, EPA used local pesticide usage and hydrogeological and meteorological conditions to model pesticide concentrations in drinking water. The agency also used information on local pest pressures to predict the types and amounts of OPs used in residential settings. The localized estimates, combined with national food exposure values, yielded realistic, probabilistic estimates of potential exposure.

Several other groups of pesticides—the *n*-methyl carbamates, the triazines, and the chloroacetalanilides—share common mechanisms of toxicity and will need to undergo cumulative risk assessments too. EPA will need to develop more sophisticated, physiologically based, pharmacokinetic models to assess exposure to the carbamate CAG. Although both OPs and carbamates inhibit cholinesterase, the effects of carbamates are much more rapidly reversed; often, no effects are evident in a few hours following low levels of exposure. Therefore, the most realistic exposure estimates should take into account the different rates at which specific carbamates are absorbed, metabolized, and excreted, as well as the temporal pattern of exposure during the course of a day.

The FQPA Children's Safety Factor: Perhaps the most controversial provision in FQPA is the paragraph directing EPA to add an extra 10-fold uncertainty factor when calculating the safe level of exposure to a pesticide. Prior to FQPA, EPA reviewed extensive animal toxicity data to determine for each study the no observed adverse effect level (NOAEL). The Agency then divided the lowest NOAEL by two uncertainty factors, a 10-fold (10×) uncertainty factor to account for the possibility that humans could be more sensitive to the toxicant than animals and another 10× to account for potential variability in humans for sensitivity to the toxicant. Thus, pre-FQPA, the safe dose for humans was generally reckoned to be 100 times lower than the amount of the pesticide that produces no adverse effects in animals. FQPA required the use of an additional 10× uncertainty factor to account for the potential that infants or children may be more sensitive than adults and for possible gaps in the toxicity or exposure databases. When the additional Children's Safety Factor is used, the safe level becomes 1/1000th of the lowest animal NOAEL.

EPA has wrestled with the application of the Children's Safety Factor. The Agency has attempted to lay out guidance for its application in multiple "science policy documents" on which EPA has sought both public comment and external scientific peer review through its SAP advisory committee. This public engagement, however, has done little to reduce the debate about the 10× provision. The controversy over the Children's Safety Factor arises in large part from the highly divergent positions taken by key stakeholders on its meaning. Environmental advocacy groups contend, in effect, that EPA does not have (and probably would never have) sufficient "reliable data" to warrant the use of any different factor. Representatives of the pesticide industry and agricultural groups argue, in contrast, that EPA should not declare that there is a data gap with respect to a pesticide, triggering the application of the additional 10× factor, until the Agency has imposed a requirement on the affected registrant and allowed the company sufficient time to conduct the required study. Practically, the industry's approach would mean that EPA would rarely, if ever, use the Children's Safety Factor. Further complicating the interpretation of the provision is the fact that Congress, in its rush to pass the FQPA, said little about the section, and what little legislative explanation exists is confusing and inaccurate in some aspects. Not surprisingly, some eight years after enactment of the FQPA the controversy over the Children's Safety Factor persists.

But, this continuing controversy actually has deeper roots. The essential tension in the Children's Safety Factor provision (and the challenge for EPA) is that this provision adopts a seemingly scientific approach to address an issue that science addresses poorly. In effect, FQPA instructs EPA how to handle uncertainty. The Agency's regulators, presumably informed by their scientists, must apply a specified safety factor to deal with uncertainty—the possibility that there may be gaps in the understanding of the toxicity of a pesticide, levels of exposure to the pesticide, or the relative sensitivity of adults and the young—unless "reliable data" allow them to use a different one.

Dealing with uncertainty involves a mixture of scientific and policy judgments. Scientists will always acknowledge that some residual uncertainty remains for every conclusion. They may even be able to characterize the level of uncertainty comparatively: for example, knowledge in one case is more "significant" or "limited" than in the other. But dealing with uncertainty is an aspect of regulatory activity that resists quantification. In contrast to variability, which science can measure and represent quantitatively, uncertainty involves the truly unknown—what result will a future test produce? At best, based on myriad factors, EPA can have a qualitative sense but can never be absolutely certain, about what new studies might likely show. The policy judgment then becomes "how to weigh accurately different types of information and what weight should be given to different areas in which knowledge may be limited."

Congress, wisely, decided that EPA does not require absolute certainty to depart from the default, additional 10× safety factor. In keeping with the overall statutory standard, Congress crafted a statute allowing EPA to use a different safety factor if "reliable" data show that the different factor is "safe," thereby incorporating further judgment about whether EPA has "a reasonable certainty of no harm." Neither of the critical adjectives—"reliable" or "reasonable"—has a generally accepted, quantifiable meaning in the field of toxicology or risk assessment. They are, rather, the kinds of words that involve some measure of subjective evaluation, i.e., a policy judgment, but nonetheless an evaluation in which weighing of data and assessment of risks must be integral.

Against this backdrop of conflicting expectations and unclear Congressional guidance, EPA has predictably engaged in a succession of case-by-case decisions that examine the weight of evidence about risks for a pesticide and has decided the course of action for each with regard to the Children's Safety Factor. The public controversy will decrease or disappear only if EPA's decisions eventually form a reasonable pattern that is recognized and accepted by the diverse stakeholders as defensible both scientifically and from a policy perspective.

IMPLEMENTING THE REQUIREMENTS FOR AN ENDOCRINE DISRUPTOR SCREENING PROGRAM (EDSP)

An endocrine disruptor has been defined as follows:

An exogenous substance or mixture that alters the function of the endocrine system and consequently causes adverse health effects in an intact organism, or its progeny, or (sub)populations (3).

The Framework for the EDSP

Faced with the challenge of implementing a law requiring EPA to set up a screening program on a cutting edge and controversial area of science, the Agency explored the formation of an outside advisory committee composed of both stakeholders and scientific experts. In anticipation of the passage of the FQPA requirements, EPA held a public scoping meeting to determine the level of interest in forming such a committee. Based on the positive response at the meeting and a survey of opinion from stakeholders, EPA announced its intent to form the Endocrine Disruptor Screening and Testing Advisory Committee (EDSTAC). EDSTAC was chartered in October 1996, and was comprised of 35 experts and stakeholders from academia, federal and state government, industry, labor, and public health and environmental groups.

At the December 1996 meeting, the Committee members quickly agreed that the scope of the Agency's mandate in the law, screening pesticides for estrogenic effects on human health, was too narrow. Members concluded that wildlife should be included because the clearest evidence for endocrine disruptors was in wildlife species. They reasoned that the scope of the screening program should also include all chemicals to which humans and the environment are exposed, that is, commercial chemicals, pharmaceuticals, cosmetics, and environmental contaminants and mixtures, because polychlorinated biphenyls (PCBs) and other nonpesticides were also known to interfere with the endocrine system. Finally, they said that the focus on estrogen was too narrow. They noted that antiandrogenic effects are extremely detrimental in developing males and superficially look like estrogenic effects; so one should look at both sex hormone systems. They also argued for the inclusion of thyroid hormone because it too has a profound effect on development, and there was direct evidence for its inclusion based on environmental studies (4).

Screening and Testing

EDSTAC and other experts agreed that a tiered approach would be necessary to efficiently evaluate chemicals for their endocrine disrupting potential and that tier 1 should be composed of both in vitro and in vivo screens (4,5). The purpose of tier 1 is to determine the potential of chemicals to interact with the endocrine system. The in vitro assays would provide useful information on mode of action such as the ability of chemicals to bind to the receptor or affect the synthesis of hormones. However, abilities of such assays are limited in that they do not factor in metabolism or the other complexities of whole biological systems. The in vivo assays would fill in those gaps and thereby identify chemicals that need to be metabolized to the active form or those that acted at other points in the hormone system such as the hypothalamic-pituitary axis which is the control mechanism for hormone synthesis and feedback.

EDSTAC stated that the tier 1 battery needed to meet five criteria to function as an effective screen:

1. It should maximize sensitivity to minimize false negatives while permitting an acceptable level of false positives.
2. It should include a range of organisms representing known or anticipated differences in metabolic activity.
3. It should detect all known modes of action for the endocrine endpoints of concern.
4. It should include a sufficient range of taxonomic groups to account for known differences in endogenous ligands, receptors, and response elements.
5. It should incorporate sufficient diversity among endpoints and assays to reach conclusions based on "weight-of-evidence" considerations.

EDSTAC reviewed current protocols and concluded that, although none were validated, the assays in Table 1 had the greater history of use and were more likely to be successfully validated and that, taken together, they would meet the criteria noted above for an effective screening battery.

The rationale for each of these assays and how they contribute to meeting the criteria was laid out by EDSTAC. The ER initiates a cascade of cell- and tissue-specific effects. Estrogen binds to the ER and forms a complex which, in combination with mRNA and other cofactors, initiates the synthesis of a protein which may be a structural protein or a functional protein (Fig. 1). An in vitro assay—either a simple receptor binding or a reporter gene assay—was therefore deemed to be an important and cost-effective assay to include in the battery to identify potential estrogenic chemicals. Similarly, an androgen receptor (AR) binding or reporter gene assay would detect

Table 1 Tier 1 Screening Assays Recommended by EDSTAC

In vitro assays
  ER binding or transcriptional activation
  AR binding or transcriptional activation
  Steroidogenesis (sliced testes)
In vivo assays
  Uterotrophic assay
  Hershberger assay
  Pubertal female
  Frog metamorphosis
  Fish screening assay

Abbreviations: AR, androgen receptor; EDSTAC, Endocrine Disruptor Screening and Testing Advisory Committee; ER, estrogen receptor.

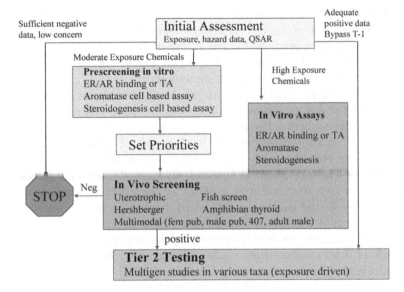

Figure 1 An alternative testing framework which incorporates in vitro prescreening. *Abbreviations*: ER, estrogen receptor; AR, androgen receptor; TA, transcriptional activation.

chemicals that mimic androgen hormones or interfere with binding to the AR. The results would be quantitative and could provide useful information about the mode of action thus helping in the interpretation of data from other assays in the battery. An in vitro steroidogenesis assay would help to identify another important mode of action: interference with the synthesis of the steroid hormones. An assay using tissue taken from the testes of rats was thought to be the best developed among the choices of assays for steroidogenesis, although concern about the ability of the assay to distinguish among chemicals that interfered with steroidogenesis from those that were merely cytotoxic was raised during discussion at Committee meetings. Both types of chemicals would lead to a decrease in measured testosterone. Distinguishing between these chemicals was thought to be the biggest challenge facing the development and validation of this assay. Unfortunately, at the time of EDSTACs report, there appeared to be no useful in vitro assay to detect chemicals that might interfere with thyroid action; however, EPA and the OECD are now investigating the progress that has been made in this area.

In vivo assays are more expensive and less specific in their response, but they complement the in vitro assays by accounting for absorption, distribution, metabolism, and excretion; responding to a broader range of mechanisms; and providing a comprehensive evaluation of the endocrine system as a unit. The uterotrophic assay was recommended for the battery to detect chemicals that are metabolized to estrogens as well as to provide for a simple evaluation of in vivo estrogenic response. The principle of the assay is that estrogen binds to the ER in the uterus of an immature or ovariectomized female rat or mouse leading to an increase in uterine tissue which is measured by an increase in uterine weight.

Similarly, the Hershberger assay is an in vivo assay to detect chemicals that interact with the AR and those that are metabolized to AR active substances. Most of the concern is for antagonists, that is, chemicals that block the AR and result in suppression of the development of normal male sex characteristics. The Hershberger is run to detect antiandrogenic chemicals in an immature or castrated male rat or mouse by coadministering a potent androgen and the test chemical and looking at the effect on androgen-dependent tissues. If the androgen-dependent tissues develop normally, one can conclude that the test substance was not antiandrogenic. It can be run to detect androgens by administering only the test chemical, in which case, the lack of development of the tissues indicates that the test chemical was not androgenic [thyroid stimulating hormone (TSH)].

The cornerstone assay in the EDSTAC tier 1 battery is the pubertal female assay. It is different from the other assays in that an intact female is used so that the interaction of all the component parts of the estrogen and thyroid hormone systems come into play. The principal estrogenic endpoint in the pubertal female assay is the effect of the chemical substance on the age of puberty as measured by vaginal opening. Effects on the thyroid system can also be assessed by measuring the thyroid prohormone, T4, and TSH.

EDSTAC was concerned about the ability of the thyroid hormone measurements in the pubertal assays to identify thyroid active substances. It therefore recommended the inclusion of a thyroid-specific assay involving metamorphosis of the frog, a process known to be highly dependent on thyroid hormone. EDSTAC also was concerned about having only mammalian assays in the battery to detect ER and AR effects. It is known, for example, that in fish 11-ketotestosterone is the most active androgen, whereas in mammals, it is dihydrotestosterone. It seemed to make sense, based on this knowledge and other information indicating that fish and mammals are at opposite ends of the vertebrate spectrum, that fish should also be included in the tier 1 battery so that substances interfering with the endocrine systems of fish, but not with that of mammals, would be detected.

EDSTAC recognized that there were some other choices for assays in tier 1 and designated them as alternative assays (Table 2). The alternative assays included the pubertal male assay, an aromatase assay, and an adult male assay. The pubertal male offered some advantages over the pubertal female in that it would probably be sensitive enough to detect chemicals interfering with steroidogenesis so that a separate assay for steroidogenesis would not be necessary. However, the pubertal male is not sufficiently

Table 2 Alternative Tier 1 Screening Assays
Recommended for Consideration

Pubertal male assay
Aromatase in vitro assay
Adult male assay

---

sensitive to inhibitors of aromatase (the enzyme that converts testosterone to estrogen); so a separate assay had to be included to detect inhibitors of aromatase if the pubertal male assay were included in the battery instead of the pubertal female assay. Proponents of the adult male assay claim that it can detect chemicals operating through all modes of action, but its critics claim that, because it is run in an adult animal, it is not as sensitive as assays in which exposure involves prenatal or prepubertal life stages.

Chemicals deemed to be positive based on a weight of the evidence in tier 1 would proceed to testing in tier 2 (Table 3). The purpose of tier 2 is to identify adverse effects and establish a dose–response relationship for these effects—information that would be required for hazard and risk assessment. It was felt that tier 2 tests would be comprised of multigeneration tests in species representing each major taxonomic group of concern. The taxonomic groups identified were mammals, birds, amphibians, fish, and invertebrates. The testing required for each chemical would be based on the expected exposure to the chemical. Thus, chemicals that are expected to enter surface water would be subject to testing in fish. If birds are expected to be exposed through water or food, testing would be required in birds, etc. EDSTAC recommended that chemicals could skip tier 1 and go directly to tier 2. The Committee suggested that this might be appropriate for food use chemicals where the multigeneration test in mammals is a part of the testing requirements for pesticide registration (40 CFR Part 158). This provision was proposed by EPA but has not yet been adopted by the Agency (6).

Priority Setting

EDSTAC recommended the screening of virtually all chemicals to which humans and the environment would be exposed. Recognizing the huge challenge this implied, the Committee recommended that the Agency develop

Table 3 Tier 2 Tests

Mammalian 2-generation reproduction and fertility assay
Fish life cycle assay
Avian 2-generation reproduction and fertility assay
Amphibian reproduction and fertility assay
Mysid 2-generation test

---

a relational database (Endocrine Disruptor Priority-Setting Database, EDPSD) containing all available information on exposure, fate, and effects, and use this tool to sort chemicals into several groups: chemicals of low concern, chemicals needing screening information, chemicals with sufficient tier 1 data indicating a need for tier 2 testing, and chemicals with sufficient data for hazard assessment. Most chemicals were expected to fall into group 2, those needing tier 1 screening data. EDSTAC recommended that the Agency set priorities for chemicals in this group and screen them in batches. EDSTAC also recognized that existing information on human and wildlife effects would generally be inadequate even for priority setting and recommended that EPA test all chemicals with an annual production volume greater than 10,000 pounds in a high throughput prescreen (HTPS). HTPS would consist of transcriptional activation (reporter gene assays) for ER and AR binding. EDSTAC pointed out that such assays were used routinely by the pharmaceutical industry to screen for synthetic hormones and hormone antagonists.

EDSP Implementation

EDSTAC report was delivered to EPA in August 1998, in time to meet the FQPA deadline for development of the EDSP. EPA accepted the EDSTAC's recommendations as the framework for the EDSP in an August 10, 1998, Federal Register notice, and formally proposed the EDSTAC's recommendations as the basis for EPA's EDSP, in December 28, 1998, for public comment.

EDSP is being implemented in three parts: assay development and validation, chemical selection (priority setting), and the development of a framework of administrative policies and procedures needed to implement the program.

Development and Validation of Assays

FQPA requires the use of validated tests. This is the greatest challenge facing EPA in the implementation of the EDSP. About the same time the FQPA was passed, the Interagency Coordinating Committee for the Validation of Alternative Methods (ICCVAM) published a report that defined validation and specified a set of procedural steps by which validation should be carried out. ICCVAM defined validation as "a scientific process designed to characterize the operational characteristics, advantages, and limitations of a test method and to demonstrate its reliability and relevance for a specific purpose" (7).

The validation procedural steps defined by ICCVAM and EPA's approach to validation are explained below. Throughout the process EPA consults with an advisory committee composed of stakeholders and experts from academia, industry, environmental groups, government, and the animal welfare community.

Test method development: Method development begins with a search of the scientific literature for information related to the test procedure. This information is captured in a detailed review paper (DRP). The DRP explains the scientific principles on which the test method rests, the purpose of the assay, and the context in which it will be used. It identifies candidate protocols and their developmental status, and compares them with respect to meeting the purpose of the assay, animal usage, cost, and other practical considerations. Finally, when possible, it recommends an initial protocol for further development or use in prevalidation.

Prevalidation: Prevalidation is a set of studies conducted to arrive at an optimized, transferable protocol. The types of studies required during prevalidation depend upon the developmental status of the test method. If a suitable test method is found in the scientific literature, a demonstration of the method is conducted. A protocol demonstration confirms the results obtained in the literature and gives the lead laboratory experience in conducting the assay. Typically, a protocol optimization study will be conducted next to choose the optimum conditions for running the assay, refine the protocol, and eliminate nonsensitive or duplicative endpoints. Special studies may also be required to address certain issues raised in the DRP or in the demonstration or optimization studies. Additional studies with the optimized protocol are conducted in the lead laboratory to obtain an estimate of intralaboratory variability, and the assay is usually run in a second laboratory to obtain initial data on protocol transferability. EPA has also decided that the multichemical study to demonstrate relevance is best conducted at this stage. It is more cost effective to conduct the multichemical studies in a single laboratory, and also better to learn that an assay is not sufficiently sensitive in tests against reference chemicals before beginning the multilaboratory validation trials. These studies will ordinarily be summarized in a prevalidation report.

Validation: The multilaboratory test phase is usually called validation. The main purpose of the multilaboratory studies is to demonstrate the transferability and reliability of the protocol. These parameters are quantified by the variability measured between different runs within a laboratory conducted at different times and between laboratories. For most validation studies in the EDSP, EPA is enlisting three to five laboratories with a relatively small number of chemicals; ideally one reference chemical per major mode of action and one or more negative chemicals would be included. Validation studies are carried out under good laboratory practices (GLPs). The validation studies from each laboratory will be summarized in a validation report which will analyze the results of the validation.

Peer review: When the laboratory work and analysis have been completed to the EPA's and its advisory committee's satisfaction, the entire body of work including the DRP is submitted to a panel of scientific experts who are completely independent of any involvement in the development and validation of the assay and who are free of any direct conflict of interest. The conditions for peer review are laid out in the ICCVAM procedures (7).

Priority Setting

Following the EDSTAC's advice, EPA built a relational database to implement what was termed "a compartment-based priority-setting approach." It was recognized by EDSTAC that the quantity and quality of information would be uneven for the majority of chemicals. To avoid biasing the selection toward chemicals with the most information, EDSTAC recommended grouping chemicals on the basis of a common data element. There would be exposure compartments, hazard compartments, and a combined compartment. Thus, chemicals within each group or compartment could be compared, and the highest scoring chemicals within each group could be selected for screening. There would be compartments based on occupational exposure, chemicals in consumer products, chemicals found in biological tissue, chemicals positive in HTPS, etc. Version 2 of the priority-setting database contained 26 compartments.

In parallel with the construction of the priority-setting database, EPA conducted a demonstration of a transcriptional activation or a reporter gene assay to determine if this technology, which was recommended by EDSTAC for use as a HTPS, could help the Agency in identifying high priority chemicals for further screening. In the demonstration that was conducted, it was found that, although the assay had been used by the pharmaceutical industry to identify highly active substances as drug candidates, it was too insensitive in its current state of development to detect most hormonally active environmental contaminants. Although most experts believed that the assay could be optimized to meet EPA's needs, the Agency thought that it needed to turn its attention from further optimization of these assays to the development and implementation of the tier 1 screens which were subject to a schedule in a Settlement Agreement with the Natural Resources Defense Counsel.[f]

In the place of HTPS, some experts urged that EPA use ER binding models to predict the chemicals that would bind to the ER. Proponents of these models pointed out that they would be faster and cheaper than testing actual chemicals; all one needed was a good mathematical model and a knowledge of the structure of the chemical. EPA embarked on validating two candidate ER-binding models. One, developed by the National Center for Toxicological Research, was known as a CoMFA model (8); the other,

---

[f]EPA was sued by the Natural Resources Defense Council for failing to implement the EDSP by August 1999 as required by the FQPA. In a Settlement Agreement with the NRDC, EPA stated that it would use best efforts to validate and implement tier 1 screens by December 2003.

developed by Dr. Ovanes Mekenyen, was known as the COREPA model (9). The ER-binding model validation program was carried out as follows: Each modeler was asked to predict the binding potential of approximately 600 chemicals. EPA selected 50 predicted to be positive by each model for testing to empirically measure its binding potential or relative binding affinity (RBA). EPA then selected approximately 200 chemicals from the 600 at random. The Agency then compared each model's prediction with the measured RBA in a standard 2×2 table (10). The results were disappointing. Some believed the results indicated that a great deal more needed to be learned about how to model the receptor–ligand interaction, specifically taking into consideration the multiple conformations the receptor–ligand complex can assume; others believed that the poor results were due to poor quality data in the models' training sets.

At this point, the Agency had neither an HTPS nor a validated binding model and thought that available data on health and environmental effects of these chemicals would be only marginally useful and not worth the cost. Thus, on December 30, 2002, the Agency proposed to pick the first 50 to 100 chemicals solely on exposure and solicited comment on this alternative (11).

International Harmonization

In 1997, the OECD established a Special Activity Endocrine Disrupter Testing and Assessment (EDTA) to provide a set of internationally recognized and harmonized test guidelines and testing and assessment strategies for regulatory use that would avoid duplication of testing and thus save resources and animals. National representatives on the EDTA met and considered approaches recommended by various expert groups including EDSTAC and the Weybridge conference. It became apparent that a common approach to screening was not going to be adopted by member countries, as the approaches of the United States, Europe, and Japan would need to reflect national requirements, and, in the case of the United States, the EDSTAC's recommendations. Nevertheless, the EDTA did agree on certain assays that member countries would use in their approach and further agreed to conduct a program to validate those test methods.

EDTA initially chose three assays to validate—the uterotrophic assay, the Hershberger assay, and a modified version of the OECD 407 assay—and formed a group to manage their validation. The United States supported the choice of the uterotrophic and Hershberger assays as these were included in the battery recommended by EDSTAC; however, the OECD 407 assay is a 28-day assay in adult animals for subacute toxicity that is not required by EPA. Despite doubts as to its sensitivity, the 407 was attractive to Japan and Europe because it was already used in testing all new chemicals and only the uterotrophic and Hershberger assays would be needed to complete the mammalian portion of the screening battery. The U.S. EPA preferred the pubertal assays, which the agency believed were more sensitive than screening assays in adult animals. The U.S. chemical industry preferred a 15-day study in the adult male rat, one of the alternative choices recommended by EDSTAC.

In the area of ecological screens, there was agreement that the fish and frog assays were necessary, but there was disagreement over what fish screening method to include. The Europeans argued for a short-term assay in adult fish whose endpoint was the production of vitellogenin, an egg protein induced by estrogen and not normally seen in male fish. The United States believed that this assay was too limited in that it would only detect estrogenic chemicals. The United States advocated a fish spawning assay; others felt this was too long and expensive for a screen and objected to fecundity, one of the endpoints of the assay, as too nonspecific for inclusion in an endocrine screening assay.

OECD assay validation program is an important adjunct to the EPA's own efforts to validate assays for two reasons: It spreads the burden of test method validation over many countries, and it will ensure international harmonization of most of the methods that will be required including all of the more expensive tier 2 tests.

Even though the goal of a common international screening and testing program proved to be unrealistic due to different national constraints—the EDTA abandoned a proposed tiered testing scheme in its draft framework in favor of a "tool box" of assays—the OECD's validation and test guideline development efforts will achieve the goal of reducing redundant testing and costs.

Effect on Regulatory Toxicology

The endocrine disruptor issue affects regulatory toxicology in three ways: it gives the EPA a new authority by which to require testing, it adds new tests to the testing requirements faced by manufacturers of chemicals and pesticide registrants and to EPA's burden to review such studies, and it may profoundly change the way chemical risks are assessed.

The Impact of a New Testing Authority: While section 408(p) of FQPA may not enable EPA to reach many additional chemicals, the discretionary authority under the SDWA to screen chemicals that may be found in sources of drinking water may enable EPA to reach by order authority many chemicals that have only been subject to the Toxic Substances Control Act (TSCA) testing authority in the past. TSCA test rules have been slow and labor intensive, at best, and with the addition of numerous additional analytical requirements for regulations by both Congress and Presidential Executive Orders, EPA has largely sought to obtain data on commercial chemicals by voluntary means. If these chemicals are found in sources of drinking water, EPA will have a new, less resource intensive way to require screens for endocrine disruption, and some of these data may also be useful in addressing other concerns.

Accommodation of Additional Testing Requirements: Initially in the United States, chemicals identified through a chemical selection process will be subjected to a battery of tier 1 screens. Contract testing facilities will have to develop the capabilities to perform these tests which may involve staff acquisition, staff training, and an expansion of testing facilities. The additional cost of the tier 1 assays recommended by EDSTAC has been estimated at $130,000 (12).

At some later stage, it seems likely that the endocrine screening and testing requirements would be incorporated into pesticide registration requirements (40 CFR Part 158). This would automatically subject new pesticides and pesticides undergoing registration review to endocrine testing. In conjunction with registration review, incorporation of the endocrine testing requirements in Part 158 would establish a 15-year cycle of testing and assessment of pesticides for endocrine effects. We can envision a period of time in which higher-priority pesticides are selected for screening through the EDSP chemical selection process while the lower-priority pesticides await their turn in the registration review timetable. Thus, over the long term, the endocrine screening of pesticide active ingredients would become an automatic, integral part of the pesticide registration process. The EDSP would then focus on pesticide inerts, drinking water contaminants, and commercial chemicals.

Given the difficulty EPA has had in amending Part 158, it would be unrealistic to expect the incorporation of endocrine testing requirements in Part 158 in the near term. However, when this does happen, how will it happen? Will these assays be add-on requirements, or can they somehow be integrated into testing requirements without increasing costs and animal usage? The mammalian and in vitro tier 1 screens are unique assays serving a specific role in the screening battery. These endpoints are not addressed in current tests and they will not contribute much information on other toxicity endpoints. Separate, ongoing efforts by the NAS and the International Life Sciences Institute to rethink how pesticides and other chemicals are screened for toxicity may lead to an approach that integrates endocrine screening with other endpoints, but one should be cautioned that combining endpoints into a single protocol could render the test too complex to conduct.

The fish reproductive screen, on the other hand, will provide useful non-endocrine information on viability and fecundity and may be capable of replacing other assays such as the fish early life stage test that is required for pesticides that raise concerns to aquatic organisms. The situation for tier 2 assays is even brighter. Although these assays are lengthy and expensive, they can be directly substituted for less complete or less robust assays currently required for food use pesticides and can be triggered for other chemicals.

Assessment of Risks: The third way in which the endocrine disruptor issue may affect toxicology is in the way risks are assessed. At this point we can only speculate because too little is known about endocrine disruption to conclude how risk assessment of endocrine disruptors will be approached.

At the present time, most risk assessments are based on the critical effect of the most toxic chemical in a mixture and the premise that a chemical either has no threshold and, therefore, always presents some finite risk (the traditional cancer risk model) or that a threshold for effects exists below which the chemical is safe (the traditional model for noncancer effects).

In the absence of contrary information, the assessment of the risk of EDCs will use the current risk assessment models. However, there are indications that future assessments of the risk of EDCs may be different from current approaches. First, it is argued that because natural hormones such as estrogen are already a risk factor for some diseases such as breast cancer and endometrial cancer, there may be no threshold for exposure to environmental estrogens (13). Some antiandrogens also have shown no threshold when tested at levels well below the previous NOAEL, although a threshold might be observed if doses were pushed low enough (14). However, it is possible that a false threshold might be observed due to failure in detecting some low dose effects due to the size of the exposed cohorts and sensitivity of the assay. Thus, in such cases one might expect that some noncancer effects caused by endocrine disruptors may be more appropriately assessed with a nonthreshold model like those used for cancer. Such assessments may drive regulatory decisions (e.g., tolerances for pesticide residues on foods, maximum contaminant levels permitted in drinking water, etc.) to lower levels if they have been based on traditional threshold models.

Beyond the threshold/nonthreshold question is the shape of the dose–response curve in low-dose regions. Most dose–response curves are monotonic, that is, as dose increases, response increases or stays the same. However, numerous in vitro studies and some in vivo studies have shown that the slopes of the dose–response curve of some endocrine disruptors are U-shaped or inverse U-shaped curves. In vivo, nonmonotonic dose–response curves can occur as a result of feedback processes controlling hormone synthesis or metabolism (15). Some chemicals cause mixed agonist/antagonist activity, and one mode may predominate at one dose level and the other mode at a different dose level. An inverse U-shaped dose–response curve would be expected from a chemical that showed weak agonist/activity at low doses and antagonistic activity at higher doses. The ER is known to exist in at least two forms, alpha and beta. The mixed agonist antagonist activity may be due to different interactions with these various forms of the receptor or it may be due to interaction with different cofactors for transcription which impose different steric requirements on the receptor–ligand complex. Chemicals can affect the endocrine system through more than one mode of action, for example, down regulation of aromatase synthesis at low concentrations and interaction with the ER at high concentrations (16). Again, the observed dose–response curve would not necessarily follow the traditional sigmoid shape but would be a composite of the two modes of action, in this case a U-shaped curve. At this time, it is not clear how prevalent or important nonmonotonic dose–response relationships are for different endpoints or how we should handle them in a testing and risk assessment paradigm.

Finally, there is the question of the effects of mixtures. Because of their complexity, the effects of mixtures are often ignored in risk assessments in favor of chemical-by-chemical assessments. However, the FQPA mandated the Agency to address the cumulative effects from multiple compounds with a common mechanism of toxicity when setting tolerances to pesticide residues in food. The focus on mechanism or mode of action allows a more sophisticated, scientifically focused approach to considering effects from multiple contaminants. Although this statute does not apply to other regulatory programs, it is changing the way the Agency is approaching hazard assessment, and what we learn here will undoubtedly affect the way we conduct hazard assessments elsewhere. Because endocrine disruption is a mode of action, we should expect this to be a prominent area for the application of this new policy. There is experimental evidence with an in vitro yeast reporter gene assay system to demonstrate that a mixture of xenoestrogens present at levels well below each of their no-observed-effects concentrations in fact produced an additive effect (17). Traditional hazard assessments would lead us to the conclusion that such mixtures are safe. Hazard assessments that take into account the additivity of chemicals that operate by a common mechanism of action will lead us to conclusions that some mixtures that we previously regarded as safe are, in fact, not safe and that will lead to more stringent control levels.

Beyond the First Priority List: Although EPA is using an exposure-only basis for the first list of chemicals subject to the EDSP, the Agency has committed to using hazard information to select chemicals for future lists. Experience to date has demonstrated that existing information is of limited use, and its review would place heavy burdens upon the Agency, the key to an approach that considers hazard information is a highly efficient tool based on mechanism. Thus, the requirement to screen numerous chemicals is driving the development of better in vitro screens and more sophisticated models. EPA will likely renew its attempts to employ HTPS and predictive models for setting priorities for screening beyond the first batch. The Agency is pursuing recombinant ER and AR assays, transcriptional activation assays for ER and AR binding, and a cell-based assay for steroidogenesis (H295R), all of which could be used to test thousands of chemicals to set further screening priorities. To complement this effort, the Agency is developing a set of biologically based models known as computational toxicology. The Agency may apply one or both of these approaches to future priority setting. One could envision the implementation of HTPS first and the use of the database of several thousand chemicals created by HTPS as the ultimate training set for the refinement of the computational toxicology models. Further refinements of the models could include correlation with in vivo results as the screening program moves forward. Robust models offer the advantage over HTPS in that no chemical procurement and testing is necessary, and predictions could guide new chemical development away from structures that might be problematic.

A new paradigm compatible with the current exposure based approach and an approach using mechanistic screens and mathematical models might look like the one in Figure 1. It would meet EPA's goal that the highest exposure chemicals should be tested to demonstrate safety regardless of suspicion, while at the same time targeting most testing resources on chemicals with a demonstrated basis for concern. As technology develops, the Agency would be expected to include other technologies such as in vitro thyroid assays. We should not be captive of the science of today but go forward as best as we can, accepting current limitations.

REFERENCES

1. Chemically-Induced Alterations in Sexual Development: The Wildlife/Human Connection. Consensus Statement of the 1991 Wingspread Conference. Available at http://www.ourstolenfuture.org/Consensus/wingspread1.htm.
2. Krimsky S. Hormonal Chaos: The Scientific and Social Origins of the Environmental Endocrine Hypothesis. Baltimore, MD: Johns Hopkins University Press, 2000.
3. Damstra T, Barlow S, Bergman A, Kavlock R, Van Der Kraak G. Global Assessment of the State-of-the-Science of Endocrine Disruptors. WHO/PCS/EDC/2002 (Online).
4. Endocrine Disruptor Screening and Testing Advisory Committee (EDSTAC) Final Report, August 1998. EPA Report No. EPA/743/R-98/003. Available at http://www.epa.gov/scipoly/oscpendo/edspoverview/finalrpt.htm.
5. European Workshop on the Impact of Endocrine Disrupters on Human Health and Wildlife. European Commission, Weybridge, U.K., December 2–4, 1996.
6. Endocrine Disruptor Screening Program. Proposed Statement of Policy. December 28, 1998. Federal Register 63:71,542–71,568.
7. National Institute of Environmental Health Sciences. Validation and regulatory acceptance of toxicological test methods: a report of the ad hoc Interagency Coordinating Committee on the validation of alternative methods. NIH Publication No. 97–3981. Research Triangle Park, NC, 1997.
8. Fang Hong, Tong Weida, Welsh WJ, Sheehan DM. QSAR models in receptor-mediated effects: the nuclear receptor superfamily. J Mol Struct: THEOCHEM 2003; 622:113–125.
9. Mekenyen OG, Kamenska V, Schmieder PK, Ankley GT, Bradbury SP. A computationally based identification algorithm for estrogen receptor ligands: Part 2. Evaluation of a hERα binding affinity model. Toxicol Sci 2000; 58:270–281.
10. Evaluation of SAR Predictions of Estrogen Binding Affinity. August 1, 2002. Unpublished study for EPA by Battelle, Columbus, Ohio.

11. Endocrine Disruptor Screening Program; Proposed Screening Approach for Initial Round of Screening; Request for Comments. December 30, 2002. Federal Register 67:79,611–79,629.
12. Borgert C. Cost Estimate Survey: Endocrine Screening Assays. Alachua, FL: Applied Pharmacology and Toxicology Inc., May 23, 2003.
13. Sheehan DM, Willingham E, Gaylor D, Bergeron JM, Crews D. No threshold dose for estradiol-induced sex reversal of turtle embryos: How little it too much? Environ Heath Perspect 1999; 107(2):155–159.
14. Gray LE Jr, Ostby J, Monosson E, Kelce WR. Environmental antiandrogens: low doses of the fungicide vinclozolin alter sexual differentiation in the male rat. Toxicol Ind Health 1999; 15:48–64.
15. Hormonally Active Agents in the Environment. Washington, D.C.. National Academy Press, 1999:110–113.
16. Almstrup K, Fernandez M, Petersen J, Olea N, Skakkebaek N, Leffers H. Environ Health Perspect 2002; 110:743–748.
17. Rajapakse M, Silva E, Kortenkamp A. Environ Health Perspect 2002; 110:917–921.

4

Chemical Risk Screening and the Regulatory Process

Margaret Johnson Wilson

U.S. Environmental Protection Agency, Office of Pollution Prevention and Toxics, Washington, D.C., U.S.A.

INFLUENCE OF THE REGULATORY PROCESS ON CHEMICAL RISK SCREENING

The Toxic Substances Control Act

Background

For purposes of discussion, chemicals can be grouped into one of three categories: pesticides, drugs, or industrial chemicals. Pesticides are specifically designed to have certain toxicological characteristics and are commonly used to exterminate pesky organisms such as roaches, termites, weeds, molds, and destructive microorganisms. Drugs are also designed to demonstrate certain biological activity that can eliminate or control infectious agents and affect or regulate biological properties, among other characteristics. Because drugs and pesticides possess inherent biological activity, there are legitimate societal concerns regarding the undesirable effects arising from human or environment exposure to these chemicals. To ensure public safety, Congress passed laws (The Food, Drug, and Cosmetic Act of 1938) for controlling drugs and (The Federal Insecticide, Fungicide, and Rodenticide Act of 1972) pesticides. These laws give the Food and Drug Administration (FDA), in the case of drugs, and the Environmental Protection Agency (EPA), in the case of pesticides, both broad and specific authority to regulate and/or control these chemicals. With the exception of items such as explosives and radiological materials, if a chemical is not a drug or pesticide it is considered an industrial chemical. This chapter will focus primarily on the role and evolution of toxicological screening as it relates to the evaluation of industrial chemicals.

Industrial chemicals are used in an incredibly broad array of products and services. Industrial chemicals are used in paints, dyes, plastics, carpets, computers, toys, aircrafts, milk and soda bottles, toothbrushes, and laundry detergents, among many other applications. The new car–smell comes from the volatile chemical components of the dashboard, seats, carpets, etc. When you walk down the aisle at the hardware store, almost any liquid on the shelves (other than pesticides) is an industrial chemical. Additionally, many of the solid materials sold are industrial chemicals or include industrial chemical components in their ingredients.

Until the mid-1970s, federal agencies lacked the authority to control industrial chemicals. There were no requirements to either test or evaluate industrial chemicals or to label or control the manufacture, use, distribution, and disposal, in an effort to protect humans or the environment from exposure to these industrial chemicals. No federal agency controlled industrial chemicals because Congress had not yet passed any legislation to provide such authority.

Over time, the scientific and technical community began to recognize that some industrial chemicals could pose risks to human health and the environment. With the growing concern surrounding polychlorinated biphenyls, among other chemicals, Congress passed a new law called the Toxic Substances Control Act (TSCA) in 1976 for the control of industrial chemicals. Within the EPA, the Office of Pollution Prevention and Toxics (OPPT), formerly the Office of Toxic Substances, was charged with the responsibility of administering TSCA.

Although TSCA gives EPA the authority to control the manufacture, import, use, distribution, and disposal of industrial chemicals, TSCA differs in very fundamental ways from the statutes of the Federal Food, Drug, and Cosmetic Act that controls drugs, and the Federal Insecticide Fungicide and Rodenticide Act that controls pesticides. Both drugs and pesticides must be registered prior to commercial use and the registration process includes extensive human health toxicity testing including tests for carcinogenicity, reproductive and developmental toxicity, chronic toxicity, neurotoxicity, mutagenicity, organ toxicity, eye irritation, skin and/or lung sensitization, and pharmacokinetics, and other endpoints. Environmental toxicity testing may also be extensive and includes testing of toxicity to fish, avian species, terrestrial animals, as well as environmental fate and persistence of the chemical under scrutiny. Testing usually involves expenditure of millions of dollars and the FDA and/or EPA typically take many years to evaluate the chemical and the associated test data prior to registration of the drug or pesticide. Testing is a prerequisite for registration and the burden of providing proof for safety rests on the proponent of manufacture, i.e., the drug or pesticide manufacturer or importer. The manufacturer of a drug or pesticide must prove to the government's satisfaction that the drug or pesticide is safe when used for the intended purpose. On approval by FDA or EPA, the drug or pesticide receives a government-approved label that clearly indicates where and how the chemical can be used. In the case of pesticides, it is a violation of law if they are used for any purpose that is not specifically described on the EPA-approved label.

In general, while the burden of providing proof for safety of drugs and pesticides rests on the manufacturer, this is not the case for industrial chemicals under TSCA. Under TSCA, there is no requirement for chemical companies to conduct testing of any kind before commencing manufacture for commercial purposes. The regulatory structure for industrial chemicals under TSCA differs dramatically from that for drugs and pesticides.

Regulating New Chemicals Under TSCA

Shortly after the passage of TSCA in 1976, EPA issued a Federal Register notice asking manufacturers and importers of industrial chemicals to notify the Agency by sending a simple postcard listing the names of all chemicals that they currently have in commerce. This list of chemicals became the Inventory of Existing Commercial Chemical Substances. In effect, TSCA inventory grandfathered the chemicals already in commerce at the time the inventory was created, because these chemicals would not be subject to the same review as chemicals not on the inventory. The original Inventory of Existing Chemicals was completed in 1979 and contained approximately 60,000 chemicals.

Under TSCA, manufacturers and/or importers are free to use chemicals on the inventory for any purpose, with requirement of registration or specific labeling. In addition, unlike for drugs or pesticides, testing of existing chemicals is not a prerequisite for their manufacture or use. However, EPA can require testing of existing chemicals if they show cause, i.e., demonstrate that there is a need for testing based on hazard and/or risk concerns. EPA must go through an expensive and lengthy process of rule making including issuing a Federal Register notice and a request for comment before requiring tests. Currently only about 10% of the industrial chemicals have any publicly available test data, and of this 10%, only a small fraction have data sets approaching the scope of data required for drug or pesticide registration.

Any chemical manufactured or imported for the first time after TSCA was enacted, and the inventory completed, was considered a new chemical. Manufacturers or importers are required to submit a standardized premanufacture notice (PMN) to OPPT for review before manufacturing or importing the chemical for commercial purposes. While drugs and pesticides must be tested extensively before manufacture, there is no requirement to conduct

testing of any kind before submission of a PMN. However, if the company
has in its possession readily available data on the chemical, the company is
required to submit these data along with their PMN submission. OPPT has
noticed that typically only 30% of PMNs contain any human health data
(endpoints vary), and only about 4.8% contain aquatic toxicity data (Vin-
cent Nabholz, EPA, OPPT, personal communication). Of those PMN submissions that do include data, usually less than 1% have the scope of data
typically submitted in support of drug or pesticide registrations. While
FDA and EPA may take many years to approve a drug or pesticide, for
new chemicals, EPA must conclude the PMN review in 90 days, although
this 90-day review period can be extended for another 90 days if necessary.

TSCA has resulted in drastically different technical review and data
requirements for new and existing chemicals. EPA can control and/or
require testing of existing chemicals, but testing and/or control can be done
only through the "notice and comment rule making" process with EPA
making the case that the chemical in question may present an unreasonable
risk to humans or the environment. However, EPA can require testing
and/or control of new chemicals if OPPT believes that the new chemical
may present an unreasonable risk. EPA has the authority to order the
testing and/or control of new chemicals and is not required to go through
the expensive and lengthy notice and comment rule making process.

Impact of TSCA's Requirements

OPPT has more than 20 years of experience in reviewing PMNs under
TSCA. While the number of PMNs received varies from year to year, OPPT
typically receives approximately 1500 PMNs annually. OPPT has received
and reviewed over 35,000 PMNs since the passage of TSCA.

As previously discussed, a typical PMN contains little toxicity information. The submitter of a PMN is required to provide information on
the chemical identity, potential impurities or byproducts, limited manufacturing data, the intended use, and maximum production volume for any
consecutive 12-month period during the first three years of manufacture.
In most cases, in the absence of human health or environmental toxicity
data, OPPT has to evaluate the potential health and ecological risks posed
by the new chemicals.

Regulatory Needs Advance the Evolution of
Computational Toxicology

Faced with the challenge of assessing a chemical in the absence of associated
data, OPPT uses the data available on structurally analogous compounds to
estimate or infer the toxicity of the untested PMN chemical. During the
initial stages of the 90-day PMN review, OPPT's technical staff evaluate
the PMN material and use literature searches, handbooks, internal data

and information, and expert judgment to identify potential analogs for use in the PMN review process. Literature on the analogs is collected, and the relevant toxicological data are extracted. An interdisciplinary team of scientists is established to assess the data. The team comprises experts in relevant areas including carcinogenicity, neurotoxicity, reproductive and developmental toxicity, mutagenicity, skin, eye, and lung irritation/sensitization, organ toxicity, and environmental toxicity. These subject matter experts then meet to evaluate available test data on the relevant analogs and identify toxicity endpoints of concern based on the identified toxicity studies. The assumption is that if the analog is structurally similar to the untested PMN material, with no other extraneous functionalities present on the molecule, the PMN material is likely to have toxicological properties and modes of action similar to that of the analog.

Using information on chemical structure, pharmacokinetics, and other factors, the subject matter experts then come to a decision regarding the untested PMN material based on the data available for the analogs. For example, the assessors may conclude that the PMN material will have a toxicity profile similar to that of the analog, or that the toxicity may be greater or less than that seen in the analog(s). If the assessors conclude that the chemical poses a low hazard, further review of the material is dropped. If a moderate or high hazard is anticipated, a potential concern is identified and the PMN is subject to further review. For PMN materials identified as presenting a potential hazard, information regarding the manufacturing, use, likely distribution routes, and potential disposal practices are generated, and this information is used to estimate potential exposures. The types of exposures estimated include general population exposures, occupational exposures, and environmental exposures. This exposure information is used, together with the hazard information, to perform a risk characterization of the PMN chemical. The result of this assessment is a quantitative or semiquantitative assessment of risk to humans and/or the environment, and the results of the risk characterization are used to determine what regulatory controls, if any, are needed. OPPT typically regulates or controls approximately 10% of PMNs submitted. The action of the Agency may take the form of required exposure controls, use controls, or other actions. The results of risk characterization guide OPPT in formulating control actions, where warranted.

Increasing Number of PMNs Submitted Annually Pushes EPA to Streamline the Assessment Process

After the original TSCA inventory was established, any new chemical entering commerce required a PMN submission. In the first two years after establishment of the inventory, relatively few PMNs were submitted to EPA. In subsequent years however, the volume of PMNs began to increase sharply. While the use of chemical analogs for the assessment of PMNs is effective, this approach is time and labor intensive. As the volume of PMN submissions increased, OPPT faced with the need to complete assessments and any associated regulatory actions in the required 90 days, needed to develop additional resources for the evaluation of PMN chemicals.

Before 1987, nearly 20% of the PMNs submitted under TSCA underwent a detailed review, which was a highly resource-intensive effort that lasted most of the mandated 90-day PMN review period. By 1987, after several years of experience in reviewing PMNs, OPPT had enough accumulated data to allow for the successful grouping of PMN chemicals into defined categories having common chemical and toxicological properties. Within these categories, it was also noted that structurally similar chemicals also shared common hazard concerns for human health and aquatic toxicity. This insight prompted the creation of OPPT's Chemical Categories Report that currently contains information on more than 50 chemical categories. These categories include chemicals for which sufficient data have been accumulated to demonstrate that hazard concerns and testing recommendations vary little from chemical to chemical within one category. When a new substance is identified as being a member of a described category, the chemical is evaluated in the context of the potential health or environmental concerns associated with that category. The Chemical Categories Report is not intended to be a comprehensive list of substances that may be subject to further action in the New Chemicals Program (NCP).

For new chemicals that fall within these defined structure-based categories, hazard concerns can more easily be identified based on the existing categorical data and EPA decisional precedents. This enabled both PMN submitters and EPA reviewers to benefit from the accumulated experimental data within EPA and allowed some reviews to be facilitated. As expected, establishing these categories has streamlined OPPT review process of many new chemical substances. Based on current information, OPPT takes action to control potential risks to human health or the environment, on approximately 10% of the PMNs submitted. Only 2% to 3% of the total number of PMNs submitted (20–30% of the above 10%) now undergo a standard review, while the remaining 7% to 8% are identified as members of a category described in the NCP Chemical Categories Report. OPPT document NCP Chemical Categories lists many of the categories that OPPT has investigated over the years (this document is available online at www.epa.gov/oppt/newchems/pubs/chemcat.htm). Due to the dynamic nature of the Internet, the URLs provided here may have been changed from the time of the writing of this document. In case a URL is no longer correct, the reader is advised to use any of the publicly available Internet search engines to locate the correct URL. Examples of the categories and information that can be found in the Chemical Categories Report include

Aliphatic amines: Members of this category can be highly toxic to all groups of freshwater organisms (i.e., fish, aquatic invertebrates, and green algae). Toxicity is related to the length of the hydrophobic carbon chains.

Cationic dyes: Cationic dyes are expected to be toxic to fish, daphnids, and algae.

Development of the Chemical Categories Report, based on many years experience of evaluating test data on analogs, allows OPPT to make hazard determinations more quickly. However, for PMN material that cannot be assigned to one of the existing chemical categories, estimations based on measured data on analogous chemicals continue to be a critical assessment approach.

The Development of Quantitative Structure Activity Relationships

The development of the Chemical Categories Report significantly advanced the hazard assessment of chemicals that lacked data. However, the steady increase in the number of new PMNs, combined with the constant need to conserve resources, resulted in a demand for greater innovation in the application of science and technology in the risk assessment process.

As the experience base of OPPT grew over the years, subject matter experts began to explore quantitative approaches that would allow for the assessment of chemicals based on chemical structure and also allow assessors to utilize experimental data currently available for existing chemicals. These interests resulted in the subsequent development and use of quantitative structure activity relationships (QSARs) for the assessment of PMN chemicals. QSARs are based on observations that the activity of a chemical can be associated with a particular structure or structural subunit of the compound. Rather than using a chemical analog to judge potential toxicity of a PMN chemical, structure activity relationship (SAR) analysis allows judgments to be made about potential toxicity based on the structural units or subunits of the actual PMN chemical. QSARs are developed based on a methodology that tries to find a mathematical relationship between a set of defined descriptors for a group of molecules and their associated property or activity. The descriptors can be functional units, shape, electrostatic field, hydrophobic propensity, or other features of the molecules. These descriptors can be derived either experimentally or computationally.

Using regression analysis, a mathematical relationship between chemical descriptors and the property or activity of the chemical can be created. The goal of the mathematical relationship is to derive an equation that can be used to predict the activity of a new compound, given only the value of the descriptors for the molecule. Many techniques can be used to derive these mathematical relationships; however, the most commonly used statistical approaches are based on linear regression techniques.

As computational toxicology advanced, QSARs that address many human health and environmental endpoints have been developed. In the absence of data, answering even the most fundamental questions about the potential hazard of a chemical can be difficult. For example, physical–chemical and environmental fate data are essential to understand the behavior of the chemical once it is introduced into the environment. Without this information, understanding the hazard potential of a chemical is difficult and estimating potential exposure can be nearly impossible. Unfortunately, many PMNs lack even this most basic information.

EPI Suite™ Estimates Physical–Chemical and Environmental Fate Properties

Development of the Estimation Programs Interface (EPI) Suite set of tools is an example of how regulatory need has helped spur the development of methods for assessing chemicals that lack data. OPPT, with significant contributions from other members of the scientific and technical community both domestic and international, developed computational estimation methods to predict physical–chemical and environmental fate properties. These individual computer-based models were integrated into the Windows-based EPI Suite. This integrated set requires the user to enter the chemical structure once, and EPI Suite will run each of the models sequentially and provide estimated physical–chemical and environmental fate properties (EPI Suite can be downloaded from www.epa.gov/oppt/exposure/docs/EPI suitedl.htm).

OPPT uses a fragment-based approach to estimate physical–chemical properties. Having compiled an extensive database of existing measured physical–chemical properties for a wide variety of chemicals, subject matter experts began to break down chemicals into individual substructural units, or fragments. Based on the available measured data, each fragment was then investigated to see if, and how, its presence in the molecule affected the property being investigated. If the presence of the fragment did affect the property, mathematical methods were used to develop a quantitative coefficient for that fragment, which describes how its presence either increases or decreases the predicted property. Once these fragment constants were established, and the corresponding coefficients determined, the final fragment-based equations could be used to estimate many physical–chemical properties of new chemicals. EPI Suite uses this fragment contribution approach to estimate melting point, boiling point, vapor pressure, Henry's Law constant, biodegradation, and atmospheric oxidation potential.

More sophisticated computational approaches have been developed, which use the fragment contribution approach described above along with additional factors to predict additional physical–chemical and environmental fate properties. These methods consider the effect of individual fragments, use chemical class-specific correction factors, and may use some of the physical properties listed above as additional input into the QSAR equations. Using this approach, endpoints such as bioconcentration factor (BCF), hydrolysis, water solubility, organic carbon adsorption coefficient, and octanol–water partition coefficient (K_{OW}) are estimated by EPI Suite.

Using these estimation methods, the development of multimedia fugacity models to estimate environmental partitioning becomes possible. Fugacity models allow assessors to estimate where in the environment a chemical is likely to be found when a steady state or equilibrium is reached. This information is very important when evaluating human and environmental exposures. For example, if results indicate that a chemical has 60% partitioning to the water, 20% partitioning to the soil, 20% partitioning to the sediment, and 0% partitioning to the air, then assessors would focus on drinking water as a source of potential human exposure and discount the likelihood of inhalation exposure in the ambient environment because the chemical is not likely to partition to air. An example output from the KOWWIN module within EPI Suite for 1,3,5-triazine 2,4-diamine,6-nonyl is shown in Figure 1. Because measured values of K_{OW} of a chemical can range from 10^{-4} to 10^8 (at least 12 orders of magnitude), its logarithm is commonly used to characterize its value. The advanced output from KOWWIN shown in Figure 1 indicates the chosen fragments and their corresponding coefficients.

Log Kow (version 1.67 estimate): 4.15

SMILES : n1c (N) nc (N) nc1CCCCCCCCC

CHEM : 2,4-DIAMINO-6-NONYL-S-TRIAZINE

CAS Num : 005921-65-3

MOL FOR: C12 H23 N5

MOL WT : 237.35

TYPE	NUM	LOGKOW FRAGMENT DESCRIPTION	COEFF	VALUE
Frag	1	-CH3 [aliphatic carbon]	0.5473	0.5473
Frag	8	- CH2- [aliphatic carbon]	0.4911	3.9288
Frag	3	Aromatic Carbon	0.2940	0.8820
Frag	3	Aromatic Nitrogen	-0.7324	-2.1972
Frag	2	-N [aliphatic N, one aromatic attach]	-0.9170	-1.8340
Factor	1	sym-Triazine ring correction	0.8856	0.8856
Factor	2	Amino triazine/pyrazine/pyrimidine correc.	0.8566	1.7132
Const		Equation Constant		0.2290

Log Kow = 4.1547

Figure 1 Sample output from the KOWWIN module of EPI Suite[TM] showing 2,4-diamino-6-nonyl-s-triazine evaluated by entering CAS RN 5921-65-3 and selecting full output option. *Abbreviation*: EPI, estimation programs interface.

Ecological Structure Activity Relationships Estimates
Aquatic Toxicity

Over the years, OPPT's experience in assessing ecological effects of industrial chemicals has increased. After many years of experience in evaluating test data for aquatic toxicity, OPPT began to draw correlations between the physical–chemical properties of a chemical and its toxicity to aquatic organisms. For example, for certain classes of industrial chemicals, a correlation was established between the K_{OW} of a material, molecular weight, and toxicity of the chemical to aquatic organisms. The mode of toxicity was assumed to be simple narcosis. These correlations were then captured in a series of regression equations, or QSARs, and in the mid-1980s, OPPT published a set of look-up tables containing the QSARs that could be used to estimate aquatic toxicity. These estimates are based on the inputs of molecular weight and the log of the K_{OW} for any chemical. These tables were entitled ecological structure activity relationships (ECOSAR). It became apparent over time that the aquatic toxicity estimates made from ECOSAR using the assumption of a simple narcosis as the mode of action, continually underpredicted the aquatic toxicity of certain chemical classes. It appeared that the materials in these classes demonstrated an excess toxicity which the regression equation for simple neutral organic chemicals did not capture. Data for these chemical classes demonstrating excess toxicity were gathered, and class-specific regression equations/QSARs that underlay ECOSAR were revised to reflect this excess toxicity.

As the capacity of desktop computers increased, OPPT automated ECOSAR, which resulted in a computerized set of QSARs. ECOSAR is available for download from the EPA Web site www.epa.gov/oppt/newchems/tools/21ecosar.htm.

Later ECOSAR and EPI Suite were integrated so that a user could either key in a measured K_{OW}, or if data were lacking, EPI Suite could calculate the K_{OW} (also called log P), and run ECOSAR automatically offering the assessor a convenient, integrated approach to estimate aquatic toxicity. The scope of ECOSAR is, for the most part, the assessment of toxicity in freshwater organisms. Regression equations are available for aquatic vertebrates (fish), invertebrates (daphnid), and green plants (algae). Six endpoints are typically estimated:

- fish acute and chronic toxicity,
- daphnid acute and chronic toxicity, and
- algal acute and chronic toxicity.

In some cases, data are insufficient to develop regression equations for all six endpoints. An example of this situation is provided below in Figure 2, that is the ECOSAR version 0.99G analysis for 1,3,5-triazine-2, 4-diamine, 6-nonyl.

SMILES : n1c (N) nc (N) nc1CCCCCCCCC
CHEM : 2,4-DIAMINO-6-NONYL-S-TRIAZINE
CAS Num: 005921-65-3
MOL FOR: C12 H23 N5
MOL WT : 237.35
Log Kow: 4.15 (KowWin estimate)
Melt Pt :
Wat Sol : 6.767 mg/L (calculated)

ECOSAR v0.99g Class(es) Found

Anilines (amino-meta)
Triazines

ECOSAR Class	Organism	Duration	End Pt	Predicted mg/L (ppm)
Neutral Organic SAR (Baseline Toxicity)	: Fish	14-day	LC50	4.273
Anilines (amino-meta)	: Fish	96-hr	LC50	1.916
Anilines (amino-meta)	: Daphnid	48-hr	LC50	0.100
Anilines (amino-meta)	: Green Algae	96-hr	EC50	0.156
Anilines (amino-meta)	: Daphnid		ChV	0.007
Triazines	: Fish	96-hr	LC50	1.676
Triazines	: Fish	14-day	LC50	4.273
Triazines	: Daphnid	48-hr	LC50	2.084
Triazines	: Daphnid	16-day	EC50	0.274
Triazines	: Fish		ChV	0.305
Triazines	: Fish (SW)	96-hr	LC50	1.086

Note: * = asterick designates: Chemical may not be soluble
enough to measure this predicted effect.
Anilines (amino-meta):
Fish and daphnid acute toxicity log Kow cutoff: 7.0
Green algal EC50 toxicity log Kow cutoff: 7.0
Chronic toxicity log Kow cutoff: 9.0
MW cutoff: 1000
Triazines :
Fish and daphnid acute toxicity log Kow cutoff: 5.0
Green algal EC50 toxicity log Kow cutoff: 6.4
Chronic toxicity log Kow cutoff: 8.0
MW cutoff: 1000

Figure 2 Sample output from ECOSAR showing 2,4-diamino-6-nonyl-s-triazine evaluated by entering CAS RN 5921-65-3. *Abbreviation*: ECOSAR, entitled ecological structure activity relationships.

Application of Expert Systems to Predicting Chemical Carcinogenicty: OncoLogic™, the Cancer Expert System

Regulatory needs demanded that OPPT develop streamlined approaches in the evaluation of carcinogenicity, when data for the chemical under concern are lacking. OPPT employed chemical analogs in the evaluation of cancer hazard potential, and in the case of chemical carcinogenicity, OPPT established a team of highly experienced toxicologists to assist in the estimation of the carcinogenicity of chemicals. EPA realized that the team specializing in chemical carcinogenesis was a very rare resource and OPPT needed to take steps to capture this high level of expertise for future use. While EPI-Suite and ECOSAR used QSAR approaches for the estimation of physical–chemical properties, environmental fate, and aquatic toxicity, related efforts indicated that an expert system was the most useful approach in predicting the carcinogenic potential of chemicals. As a result, the EPA team of carcinogenicity experts set about the task of building a Cancer Expert System designed to mimic the judgment of highly qualified experts and predict chemical carcinogenicity based on analysis of a chemical.

The EPA team analyzed the results of all relevant cancer bioassays and used insights gained from these efforts to develop numerous knowledge rules. These knowledge rules were then integrated, tested, evaluated, and ultimately computerized to yield a Cancer Expert System known as Onco-Logic™. This system is organized into 48 chemical classes. The user selects the appropriate chemical class and draws the structure of the chemical to be evaluated using a drawing program that is integrated into the system. The system then analyzes the chemical structure, employing the automated decision tree, and renders a judgment regarding the carcinogenicity potential of the chemical.

The results provided by OncoLogic include a semiquantitative estimate of cancer hazard potential of the chemical. Possible results are classified as low, marginal, low-moderate, moderate, high-moderate, and high. OPPT is making OncoLog pubicly available at no cost.

In addition to offering a semiquantitative estimate of cancer hazard potential, the Cancer Expert System also offers the rationale for the estimation provided. Because the Cancer Expert System is based on the chemical structure, the user can make subtle changes in structure and observe the associated change in carcinogenic hazard potential, if any. In Figure 3 was shown the output of OncoLogic for the chemical 1,3,5-triazine-2,4-diamine,6-nonyl.

In addition to EPI Suite, ECOSAR, and OncoLogic, OPPT has developed other computerized methods that are useful in chemical-risk screening. These methods include tools for estimating human exposures, both general population and occupational, as well as exposure of the aquatic environment. OPPT uses these tools to conduct screening-level risk assessments of chemicals in the absence of data.

SMILES : n1c(N)nc(N)nc1CCCCCCCCC
CHEM : 2,4-DIAMINO-6-NONYL-S-TRIAZINE

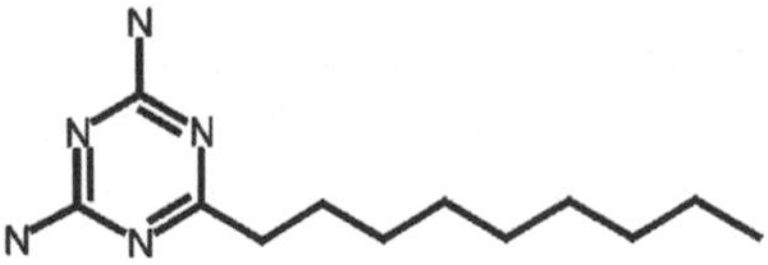

CAS Num: 005921-65-3

The level of carcinogenicity concern for this compound is LOW.

JUSTIFICATION.

In general, the level of carcinogenicity concern of an aromatic amine is
determined by considering the number of rings, the presence or absence of
heteroatoms in the rings; the number and position of amino groups; the nature,
number and position of other nitrogen-containing 'amine-generating groups';
and the type, number and position of additional substituents.
Aromatic amine compounds are expected to be metabolized to N-hydroxylated/N-
acetylated derivatives which are subject to further bioactivation, producing
electrophilic reactive intermediates that are capable of interaction with
cellular nucleophiles (such as DNA) to initiate carcinogenesis.
An aromatic amine containing one 6-membered ring with three carbon ring atoms
replaced with nitrogens heteroatoms, where the nitrogen heteroatoms are all
meta-to one another, two amino groups, and one alkyl or alkoxy (C>2) group
has a carcinogenicity concern of LOW.

The final level of carcinogenicity concern for this compound is LOW

Figure 3 Sample output from the Cancer Expert System showing 2,4-diamino-
6-nonyl-s-triazine evaluated by replying to questions about the chemical.

Dealing with the Inherent Uncertainty of Screening Level Methods

Previously described screening-level assessment methods developed by OPPT
to evaluate PMNs have inherent uncertainties. EPA recognizes this and to protect human health and the environment uses scientifically sound methods to
address the uncertainties. Several of these approaches employed by OPPT in
dealing with the inherent uncertainty of screening-level methods are described
in this section. This summary is extracted from a manuscript developed by
OPPT senior scientist, J.V. Nabholz (Toxicity Assessment, Risk Assessment,
and Risk Management of Chemicals Under TSCA in United States).

Conservative Assumptions

A critical approach that the Agency always uses when dealing with uncertainties
is to make conservative assumptions in the absence of data to demonstrate
otherwise. One example of a conservative assumption made by OPPT when
predicting toxicity of PMN materials in the absence of data is a PMN submission in which the reaction product has a chemical name indicating that varying
carbon chain lengths are possible. OPPT always assumes that the most toxic
structure of the structures possible according to the name may be the product

synthesized, and makes predictions based on that structure. Another example of conservative assumption made by OPPT in predicting toxicity of chemicals that lack data is assuming the presence of 100% active ingredients in the product to which humans or the environment are exposed.

Margins of Exposure

Margins of exposure (MOE) are used as a scale of protectiveness of human health. The MOE is calculated by dividing the toxicity endpoint by the predicted exposure. Higher MOE indicate lower risk. If the assessment of risk is based on a no observed adverse-effect level for the toxic endpoint of concern, an acceptable MOE is greater than 100. If risk estimation is based on a lowest observed adverse-effect level, an acceptable MOE is greater than 1000.

Assessment Factors

Assessment factors (AsFs) are used by OPPT in an effort to deal with the uncertainty associated with the extrapolation of screening-level estimates of toxic effects to individual organisms (e.g., fish) to the assumed integrity of the entire ecosystem (the stream or river). Four AsFs were described by EPA in 1984: 1000, 100, 10, and 1 [United States Environmental Protection Agency (USEPA) 1984. Estimating "concern levels" for concentrations of chemical substances in the environment]. The four AsFs are used to extrapolate from the species level to the ecosystem level under the following conditions:

- AsF 1000 is used when only one acute toxicity value is available.
- AsF 100 is used when acute values are available from several species (extrapolate using the most sensitive species).
- AsF 10 is used when a chronic value from the most sensitive group is available (fish, invertebrates, and green algae).
- AsF 1 is used (essentially no uncertainty is expected) when test data are available from micro- or mesocosm studies, field studies, or from the natural environment.

In practice, OPPT's new chemical assessments for ecotoxicity use an assessment factor of 10, as OPPT predicts chronic values for each new chemical via SAR.

TSCA 25 YEARS LATER: RISK SCREENING BEGINS TO DRIVE REGULATORY POLICY

After TSCA was enacted, priorities in regulation helped the development of methods to predict toxicological properties of industrial chemicals. Significant advances were made in the development and application of QSARs, expert systems, and other computational approaches in the evaluation of chemicals in the absence of data. Other factors began to influence advancements in predictive toxicology, and helped increase the importance of predictive toxicology to regulatory concerns. Two of these factors were particularly influential in this process: (*i*) the Pollution Prevention Act and (*ii*) several key industry–Agency collaborations designed to evaluate and articulate the advantages of the application of predictive modeling approaches in chemical product development.

The Pollution Prevention Act

In 1990, Congress passed the Pollution Prevention Act, and in so doing, created a bold national objective for environmental protection by describing a hierarchical approach for dealing with pollution.

- First: Pollution should be prevented or reduced at the source whenever feasible; pollution that cannot be prevented should be recycled in an environmentally safe manner whenever feasible.
- Second: Pollution that cannot be prevented or recycled should be treated in an environmentally safe manner whenever feasible.
- Third: Disposal or other releases into the environment should be employed only as a last resort and should be conducted in an environmentally safe manner.

Pollution prevention (P2) means source reduction, as defined under the Pollution Prevention Act. The Pollution Prevention Act defines source reduction to mean any practice which

- reduces the amount of any hazardous substance, pollutant, or contaminant entering any waste stream or otherwise released into the environment prior to recycling, treatment, or disposal, and
- reduces the hazards to public health and the environment, associated with the release of such substances, pollutants, or contaminants.

To employ the provisions of the Pollution Prevention Act, OPPT began to re-evaluate how advances in the chemical-risk screening of industrial chemicals could be applied toward P2 objectives established by the Pollution Prevention Act.

Understanding the Problem Is Key to Identifying P2 Opportunities

Each year, the chemical industry develops thousands of new chemical substances—chemicals previously unknown in the market. In many cases, when alternative chemicals or processes are considered at research and development (R&D), factors such as efficacy, yield, performance, and cost are weighed before decisions are made on which alternative to commercialize. EPA receives approximately 1500 PMNs per year; however, industry has made thousands of other decisions early in the R&D process long before chemicals are chosen and submitted as PMNs. By the time EPA starts checking a PMN, most of the opportunities to prevent pollution are lost. If a company has readily ascertainable data on a PMN chemical, under TSCA, this information must be submitted with the PMN. Only a fraction of PMN submissions have data and a much smaller fraction of those that actually have data have the kinds of data typically required to be submitted for a drug or pesticide. As a result, in some cases, industry has made commercialization decisions without understanding risk tradeoffs of products or process alternatives.

Information Informs Decision-Making

To identify and take advantage of P2 opportunities, companies and other stakeholders need access to risk-related information early in the R&D process. Companies often decide which chemicals or processes to use primarily on the basis of cost, performance, and other criteria. If companies had access to risk-related information about chemicals, they could improve their decision-making and take advantage of P2 opportunities. To address this issue, OPPT began to evaluate the predictive methods developed within EPA to support new chemical review under TSCA. OPPT wanted to know if these computational methods could be used by the industry to evaluate chemicals at R&D. OPPT believed that if the industry used these methods at R&D, this would allow companies to compare and contrast product and process alternatives based on risk-related considerations. Information generated by these methods could help facilitate the decision process employed by developers of new chemicals thus leading to the identification of many more P2 opportunities and the commercialization of inherently safer chemicals.

Eastman Kodak Pilot Project Begins

OPPT wanted to learn if the risk-screening methods developed and used by the Agency to screen new chemical submissions could be successfully transferred to the chemical industry. To address this question, OPPT approached a wide array of chemical companies, seeking partnership for a pilot project with the goal of learning if OPPT's computational screening methods could be successfully transferred to the industry. Eastman Kodak, a major manufacturer of photographic materials, demonstrated leadership and stepped forward to collaborate with OPPT in this effort.

During the meetings, OPPT staff worked with Kodak to help them learn how to use the chemical assessment methods, correctly interpret results, and understand the limitations of the methods.

Kodak and OPPT then worked together on a series of test cases. Kodak provided a set of chemicals to OPPT and both Kodak and OPPT independently evaluated the chemicals using OPPT's assessment methods. Kodak and EPA then met to compare assessments on each of the Kodak test chemicals. The results generated from each of the models from Kodak and OPPT's evaluation were nearly identical. Kodak has continued to use these assessment methods and has shown leadership in promoting their use within the chemical industry.

The P2 Framework Documents OPPT's Screening Models

Based on the highly successful Kodak pilot project, OPPT integrated all the chemical assessment tools into a program called the P2 Framework. The P2 Framework integrates all of OPPT's chemical assessment methods into a stand-alone training program. OPPT then developed an aggressive outreach plan offering seminars and workshops on the use, interpretation, and limitations of the P2 Framework. OPPT's goal was to encourage companies to use the P2 Framework at R&D, to identify problematic chemicals early on in R&D, thus leading to commercialization of inherently safer products and processes. It is important to note that OPPT's P2 Framework is a set of screening methods, and like any other screening approach, the P2 Framework has its own limitations. The training programs that OPPT offered on the P2 Framework clearly articulates these limitations. The P2 Framework is available online at www.epa.gov/oppt/p2framework/.

Industry Agency Collaborations Evaluating and Articulating the Advantages of Using SARS and QSARS in Chemical Product Development

Project XL Offers Regulatory Relief for P2 Innovation

Shortly after OPPT released the P2 Framework, EPA launched an initiative aimed at improving approaches to environmental protection. This initiative, Project XL, was open to all including chemical companies, utilities, other manufacturing facilities, etc. The objective was to invite stakeholders to come forward with better, innovative ideas and approaches toward environmental protection. Companies and organizations, interested in participating in this effort submitted proposals to EPA and if the proposal was accepted, the participating organization qualified for regulatory relief. Two companies, Eastman Kodak and PPG Industries, submitted XL proposals that were based on OPPT's P2 Framework.

Eastman Kodak Project XL Demonstrates Business Benefits of Prescreening Chemicals

Kodak submitted a Project XL proposal based on the use of OPPT's P2 Framework screening models. For many years, Kodak has had a robust chemical assessment process. In Kodak's proposal, the company described its ongoing chemical assessment approach, including the use of specially trained subject experts. In its Project XL proposal Kodak agreed to

- incorporate and use the P2 Framework in product development efforts,
- share its experience in using P2 Framework with others in the industry through publications, presentations at conferences and meetings, etc., and
- conduct a study to articulate the economic and business benefits of applying risk-screening early on in R&D.

Kodak very effectively addressed the first two commitments by building the P2 Framework into Kodak's R&D efforts and by writing articles, giving talks, etc., about the benefits Kodak accrued by using the P2 Framework. To address the economic and business benefits of risk-screening, Kodak partnered with the Tellus Institute, a Boston-based firm specializing in environmental cost accounting. The Tellus Institute worked closely with Kodak to describe the approach, and to compare the costs of new product development, both before and after Kodak adopted the P2 Framework. The study concluded that use of the P2 Framework

- reduced product development costs by 13% to 100%,
- decreased waste generation,
- allowed Kodak to evaluate more product alternatives in less time and at lower cost,
- enhanced the contributions of the health, safety, and environment staff,
- helped Kodak develop environmentally preferable products and processes, and
- decreased regulatory uncertainty.

Companies often spend considerable resources in the R&D stage of product development. If, after submitting the subsequent PMN to EPA, the chemical is severely regulated, all the product development resources invested in that product may be lost and cannot be recovered. Kodak's work in collaboration with the Tellus Institute demonstrated that risk-screening at R&D has both strong environmental and business benefits. The Tellus Institute report included several case studies of benefits Kodak accrued as a result of use of the P2 Framework at R&D. In one case study, Kodak saved $750,000 to $1,000,000 and got to market 1.5 to 2 years earlier than would have been the case if the company had not used the P2 Framework. A copy of the Tellus Institute report, entitled Design for Competitive Advantage: The Business Benefits of the EPA P2 Assessment Framework in New Product Development, is available online at www.epa.gov/oppt/p2framework/docs/p2pays.htm.

PPG Industries Project XL Validating Aquatic Toxicity QSARs

PPG Industries, a major manufacturer of paints, coatings, glass, and fine chemicals, also submitted a Project XL proposal based on the P2 Framework. As was the case with Kodak, PPG industries also agreed to

- incorporate and use the P2 Framework in product development efforts, including incorporating the assessment methods into PPG's Gate Keeper Chemical Evaluation System, and
- share PPG's experience in using the P2 Framework with others in the industry through publications, presentations at conferences and meetings, etc.

PPG very effectively incorporated the P2 Framework into the company's Gate Keeper System that ensures that health and safety considerations are addressed during product development. PPG also very effectively reached out to others in the chemical industry regarding the benefits of using the P2 Framework at the R&D stage of new chemical development. PPG made presentations at scientific meetings and conferences, among other venues, and has been an effective proponent for incorporating risk screening and P2 principles into product development.

In addition, PPG conducted an independent validation of the components of the P2 Framework through the ECOSAR program for estimating aquatic toxicity. PPG had submitted several dozen chemicals to OPPT under the PMN program. Subsequently, PPG submitted the same chemicals to Environment Canada. While the United States does not require up-front testing, Canada does require that testing including one for aquatic toxicity be conducted, and that the data be submitted with the new chemical notification.

PPG compared the aquatic toxicity test data to the ECOSAR predictions and concluded the predictive capability of ECOSAR was 87% to 90% for the same substance (Comparison of Aquatic Toxicity Experimental Data with EPA/OPPT/SAR Prediction on PPG Polymers, J.S. Chun, J.V. Nabholz, and M.J. Wilson. PPG Industries, Inc., Pittsburgh, PA, and United States Environment Protection Agency, OPPT, Washington, D.C.). ECOSAR estimations were considered predictive if the estimate was within one order of magnitude of test data. Accuracy within one order of magnitude is considered highly predictive when evaluating QSAR methodology.

Regulatory Relief: A Secondary Benefit

In their XL proposals, both Kodak and PPG requested regulatory relief. In particular, both firms asked that the customary 90-day review period be shortened to 45 days for new chemical submissions developed under their respective Project XL proposals. OPPT agreed and, changing administrative policy regarding simultaneous submission of test market exemptions and PMNs, allowed both Kodak and PPG to go to manufacture in 45 days.

Both Kodak and PPG participated in Project XL for three years. Both firms submit many PMNs annually. It is interesting to note that, though they qualify for regulatory relief, Kodak has never exercised the option. The Agency has learned that, while regulatory relief is appealing, perhaps the most significant benefits may be reduced product development costs, decreased generation of waste, reduced regulatory liability, and the ability to deliver a product to the customer on schedule.

Sustainable Futures Evolves from Successful Project XLs

Chemical Risk Screening Begins to Impact Regulatory Policy

The Agency was very pleased with the P2, risk reduction, and waste reduction benefits seen in the Kodak and PPG XL Projects and, consequently, decided to scale-up this effort nationally, making the associated benefits of prescreen available to all companies that wished to participate. Toward this end, OPPT launched a new initiative called Sustainable Futures. Sustainable Futures is the programmatic structure OPPT uses to make the P2 Framework risk-screening methods broadly available to the chemical industry. Under Sustainable Futures OPPT offers participating companies

- the P2 Framework computerized chemical risk-screening methods,
- training workshops in the use, applicability, interpretation, and limitations of the P2 Framework,
- detailed, company-specific, one-on-one training,
- a small business assistance plan, and
- regulatory relief for qualifying low hazard and/or low risk PMN submissions.

Sustainable Futures is an effective forum for the Agency and industry to work collaboratively toward our shared goals of P2, risk reduction, stewardship, and sustainable chemical use.

We Have Come Full Circle

In the early years, the regulatory challenges of implementing TSCA necessitated dramatic changes in the application of computational toxicology for the evaluation of chemicals. These regulatory challenges were the driving force behind the development of SARs, QSARs, Expert Systems, and other methods for the estimation of human and environmental risk. These methods evolved partially in response to the increasing need for rapid, computational approaches to chemical risk screening. The availability and broad use of these computational approaches is having the effect of bringing risk screening to predict toxicity and regulatory policy full circle. Now the availability of a wider range of increasingly sophisticated computational toxicological methods is helping drive changes in regulatory policy, i.e., regulatory relief, as can be seen in the Sustainable Futures Pilot Project. This trend is likely to continue. The persistence, bioconcentration, aquatic toxicity (PBT) chemical profiler is a case in point.

The PBT Profiler: Regulatory Priorities Drive Chemical Risk Screening Online

The recent past has seen dramatic advances in the capacity of personal computers. These advances, coupled with equally dramatic advances in computational toxicology, combine to make chemical screening, with its associated risk reduction and P2 benefits, broadly available. With the advent of the Internet and the World Wide Web, it was just a matter of time before computational toxicology went online in real time.

The PBT Profiler: Chemical Risk Screening Brings EPA, Industry, and NGOs Together

The PBT profiler is a no-cost, web-based chemical screening tool intended to help in the identification of persistent, bioaccumulative toxic chemicals, based on a computational analysis using chemical structure.

There is general agreement among the scientific and regulatory community that chemicals of greatest interest are those that persist in the environment for long periods of time, and bioconcentrate in living organisms and present toxicity profiles of concern, i.e., PBTs. To address this concern from a regulatory perspective, OPPT issued regulations requiring emission reporting under the Toxic Release Inventory (TRI) for chemicals that were known PBTs. In addition, EPA's OPPT issued a policy statement indicating that new chemicals that met certain PBT criteria would be subject to regulation. While the Agency knew certain chemicals had PBT characteristics of concern (DDT, PCBs, dioxins, toxaphene, among others), the vast majority of industrial chemicals have never been tested for PBT characteristics. Companies and consumers alike are generally unaware of the PBT characteristics of chemicals that are used commercially. Walking down the aisle of a hardware store, anything liquid is a chemical—which of these chemicals, might be PBTs?

The Agency, industry, and nongovernmental organizations alike recognized the need for a tool that could quickly screen chemicals for PBT characteristics and flag those chemicals that pose a hazard concern for all three endpoints. Toward this end, EPA/OPPT, together with the major U.S. chemical trade associations (The American Chemistry Council and The Synthetic Organic Chemical Manufacturers Associations) and additionally with a number of leading chemical manufacturers and the assistance of Environmental Defense, collaborated in the development of the PBT Profiler.

Table 1 Persistence, Bioaccumulation, and Toxicity Criteria Established by the U.S. EPA

Property and endpoint	Low concern	Moderate concern	High concern
Environmental persistence in water, soil, sediment	<60 days	*>60 days*	**>180 days**
Environmental persistence in air	≤2 days (not a concern)	*>2 days (is a concern)*	
Bioaccumulation fish BCF	<1000	*≥1,000*	**>5000**
Toxicity fish chronic toxicity	>10 mg/L Or no effects at saturation	*0.1–10 mg/L*	**<0.1 mg/L**

Abbreviations: BCF, bioconcentration factor; PBT, persistence, bioconcentration, aquatic toxicity.

The PBT Profiler employs methods included in EPI Suite and ECO-SAR to screen chemicals for potential PBT characteristics. The PBT Profiler estimates persistence, bioconcentration potential, and aquatic toxicity and compares these estimates to PBT regulatory criteria under TRI and the New Chemical PBT Policy Statement. As seen in Table 1, there are two sets of criteria, i.e., the concern criteria (italics) and the high concern criteria (bold and underlined). Criteria shown in regular font indicate the chemical that presents a low concern. When shown in color, these results are (i) low concern, green (shown in this text as regular font); (ii) moderate concern, orange (shown in this text as italics); and (iii) high concern, red (shown in this text as bold and underlined).

The PBT Profiler results consist of a one-page fact sheet, as shown in Figure 4. The PBT Profiler provides three levels of output. The first level of output, found on the one-page fact sheet, is a simple color-coded P, B, and T determination for the chemical under review. Results from the PBT Profiler are best printed on a color printer. However, if a black-and-white printer is used, a black-and-white version of the results is available. If the chemical exceeded the first level criteria the letter(s) appear orange, or in black and white, as italics. For example, if persistence exceeds 60 days, but is less than 180 days, the letter P appears orange or italics. If persistence exceeds the high-concern criteria, i.e., 180 days, the letter P appears red or bold and underlined. In order to be a presumptive PBT, all three letters, i.e., P and B and T, must be orange (italics), red (bold and underlined), or a combination of orange (italics), and red (bold and underlined). If any parameter, P, B, or T, does not exceed the criteria, the letter appears green (normal font in black and white). If any of the three letters, P, B, or T, is green, screening indicates that this chemical may not have PBT concerns.

The second level of output describes environmental partitioning (percent of the chemical in each media at equilibrium) and provides

Methodology . Criteria . Definitions . Chemicals That Should Not be Profiled

Home . Start a New Profile . Result . Terms of use . Security

Results

Italicized or underlined highlights indicates indicate that the EPA criteria have been exceeded.
Color version

	Persistence			Bioaccumulation Toxicity

5921 65 3 2, 4 DIAMINO 6 NONYL-S-TRIAZINE

PBT Profiler Estimate = *PBT*

Media	Half-Life (days)	Percent in Each Medium	BCF	Fish ChV (mg/1)
Water	38	■ 11%	15	*0.3*
Soil	75	■■■■■ 85%		
Sediment	340	▮ 4%		
Air	1.4	0%		

P2 Considerations and more information

Figure 4 Sample PBT profiler results showing 2,4-diamino-6-nonyl-s-triazine evaluated by entering CAS RN 5921-65-3. *Abbreviation*: PBT, persistence, bioconcentration, aquatic toxicity.

quantitative estimates for P, i.e., half-life, B, and T. Certain classes of chemicals containing functional groups associated with mammalian toxicity are flagged as well. The PBT Profiler has a substructure searching function, which determines if the chemical being screened has any substructure presenting human health concerns, as described in OPPT's Chemical Categories Report (described previously).

The third level of output, the P2 considerations pages, provides a variety of information useful for risk management and for identifying P2 opportunities. This page also indicates if measured data were used or if the properties were calculated, among other information.

The PBT Profiler calculates PBT characteristics from the structure of the chemical, and the user must enter the structure of the chemical to run the model. For existing chemicals, the user can enter the Chemical Abstract Service (CAS) Registry Number, and if the chemical is in the accompanying look-up database, the structure is retrieved and entered into the model as a Simplified Molecular Input Line Entry System (SMILES) notation. For chemicals that are not in the look-up database, the user can either enter the SMILES notation,[a] or draw the structure using the drawing program built into the PBT Profiler. Generally, the PBT Profiler calculates physical–chemical properties needed to estimate PBT characteristics. The PBT Profiler is also linked to a database of measured physical–chemical properties, and if experimental data are available, these data will be imported and used in subsequent calculations. If measured data are used, it is indicated on the P2 considerations page. The PBT Profiler model is available at www.pbtprofiler.net.

A quarter of a century ago, when Congress passed TSCA, regulatory priorities drove the evolution of chemical risk screening as it relates to the assessment of industrial chemicals. Now, 25 years later, we see chemical risk screening influencing regulatory policy, with regulatory incentives offered to those who apply advancements in computational toxicology toward the development of safer chemicals. Chemical risk screening has come of age, evolved to respond to regulatory necessities, ventured onto the World Wide Web, and is helping shape regulatory policy. Advances in computer technology, computational toxicology, and chemical risk screening are enabling scientists to work smarter and more efficiently while enhancing our abilities to safeguard human health and the environment.

---

[a] SMILES, a Chemical Language and Information System. D. Weininger, 1. Introduction to Methodology and Encoding Rules, Medicinal Chemistry Project, Pomona College, Claremont, California, 1987.

5

The Influence of Regulation on Toxicology

Mamata De, Owen McMaster, and Wendelyn J. Schmidt
Center for Drug Evaluation and Research, U.S. Food and Drug Administration, Silver Spring, Maryland, U.S.A.

In 1903, Dr. Harvey Wiley established a volunteer "poison squad" of young men who agreed to eat only foods treated with measured amounts of chemical preservatives, with the object of demonstrating whether these ingredients were injurious to health. Chemicals fed to the young men included borax, salicylic, sulfurous, and benzoic acids; and formaldehyde. (FDA Consumer: the story of the laws behind the labels.)

INTRODUCTION

Historically, the regulations governing the drug approval process have been driven by health crises. The devastating effects of the elixir of sulfanilamide and thalidomide led to major changes in the way drugs were regulated in the United States. However, as the story of Dr. Wiley's Poison Squad illustrates, science is now playing a larger role, and initiatives such as the International Conference on Harmonization (ICH) seek to standardize the process worldwide. More recently, economics and the need for timeliness have also played a larger role in the Prescription Drug User Fee Act (PDUFA).

The authority of the Food and Drug Administration (FDA) is determined and controlled by the Congress. The major mission of the drug arm of the FDA (Center for Drug Evaluation and Research or CDER) is to determine if drugs are safe and effective. FDA has no control over the pricing of drugs. Vitamins and traditional medicines are also outside the purview of the FDA.

History

In 1906, the state of the nation's food processing industry was exposed in Upton Sinclair's novel, *The Jungle*. Although the book focused primarily on meat packing in turn-of-the-century immigrant Chicago, truth in labeling of drugs was affected. Congress passed the Pure Food and Drug Act of 1906 to safeguard the purity of foods and patent medicines. Products were required to be clearly labeled indicating their contents. At this time, no requirements for testing safety or efficacy were considered.

More than 30 years later, in 1938, sulfanilamide (containing diethylene glycol) caused more than 100 deaths, mostly in children. As a result, Congress enacted the Federal Food, Drug and Cosmetic Act of 1938. This legislation required the testing of new drugs to ensure their safety prior to marketing. The data would be submitted to the FDA for review under a New Drug Application (NDA).

The next major crisis (and accompanying legislation) was in 1962 following the thalidomide disaster. Thalidomide, *a sleep agent*, was approved for use in Europe while it was still being reviewed in the United States by Dr. Frances Kelsey. Reports of severe limb defects in newborns alerted her to the teratogenicity of thalidomide and this stopped the approval of the drug. Subsequent congressional action, the Kefauver Harris Amendment, required that drugs be proven both safe and effective prior to full marketing approval.

The amendment also required that the FDA be given full details of the clinical investigations, and the animal data that support the clinical trials. The investigational new drug (IND) package was born. The statutes established time frames for the review of the nonclinical data prior to commencing the initial clinical trials: 30 calendar days to determine if it is safe to commence clinical trials using the proposed patient population, schedule, dosage, and duration.

The PDUFA of 1992 reflected concerns over how drugs are approved rather than the safety and efficacy of the drugs. In the 1970s and 1980s, the time taken to approve a drug after the submission of an NDA had slowly increased. Reasons for the increased time on the FDA side included insufficient staff, outdated information storage and retrieval, and conflicting priorities for reviewers, not to mention the sheer volume of data. Requests for clarification of data or more data from the industry could result in long lags in the review process as the companies conducted studies or put together the required data. PDUFA allowed the FDA to charge a flat fee for each NDA submitted for licensing in exchange for more predictable review times. The money was used to hire extra staff and update the computer and IT systems. An action on the NDA had to be taken within six months for a priority drug (e.g., to treat life-threatening diseases such as cancer, diabetes, AIDS) and 12 months for standard applications. The renewal of PDUFA in 1997 added further responsibilities and deadlines for reviews of special protocols. For toxicologists, this included concurrence on the doses and protocols for two-year rodent carcinogenicity studies. Review times on nonpriority drugs were shortened to 10 months.

Within the goals of achieving safe and effective drugs, the requirements for filing are laid out in the Code of Federal Regulations (CFR). Good Manufacturing Practices (GMP) for the chemistry sections, Good Laboratory Practices (GLP) for animal data, and Good Clinical Practices (GCP) are all discussed. Even the format for the final package insert label is detailed. The usual procedure is to file an IND with the initial animal studies to support the human clinical protocol. Phase I studies to determine the toxicity (no benefit expected) are then conducted. A Phase II meeting may be held between the agency and the sponsor to discuss doses, inclusion criteria, design, or other aspects of the Phase II efficacy studies. With the results of this study in hand, another (end of Phase II) meeting may be held with the FDA to aid in design of the definitive Phase III efficacy studies. Pre-NDA meetings deal with the formats (including electronic submissions) for the NDA. Multiple meetings during the NDA review cycle may be held between the FDA and the sponsor, particularly to negotiate the final package insert label. Occasionally, with complex applications, a new drug class, or questionable risk–benefit ratios, the data may also be brought before an outside Advisory Panel of experts in the particular drug area. The decisions of the panel are generally followed but are not binding.

Special circumstances have changed the way in which drugs are handled during the approval process. In the late 1980s, with the rise of AIDS activists, there were pressures to make the drugs available prior to its approval, speed the approval, and expand access for patients in clinical trials. These pressures resulted in the accelerated approval and treatment protocols. Accelerated approval is used only with drugs for a life-threatening indication serving an "unmet need." This allowed the clinical trials to proceed using a surrogate marker associated with the ultimate measure (such as CD4 counts or HIV viral loads rather than mortality from HIV infection). A single Phase III study (or sometimes even a Phase II study if sufficiently compelling) would be sufficient for initial approval; however, a second study would have to be conducted postapproval using a nonsurrogate endpoint (Phase IV commitment). Treatment protocols allow for patients meeting a set inclusion criteria to enter into an ongoing clinical trial while the drug is undergoing FDA review.

Requirements across the international drug scene have not always been consistent. As drug companies merged to become truly global, seeking approval in Europe, Japan, and the Unites States was not only desirable, but a single format for submission of the NDA in each jurisdiction also would be attractive. The ICH was initiated in 1991 to arrive at a single document and dataset for all three geographic areas. Chemistry, clinical, and toxicology issues have been dealt with in these discussions. The individual toxicology guidances arising from these negotiations will be discussed later in this chapter.

There is an established process for the issuance of guidance documents. First, the need for more information on a topic that crosses divisions in the CDER is established. The aim is to clarify the science and process which the agency uses in forming decisions based on the current science. The document is drafted by either an individual or a committee. Comments on the draft are made by the Pharmacology/Toxicology Coordinating Committee (PTCC), which primarily consists of all the pharmacology/toxicology supervisors in CDER. Once the draft is approved by the PTCC and revised by the authors, it then goes to the office directors in CDER. The next version is sent to the Federal Register for comments from the industry and other interested individuals. After incorporation of relevant outside comments, the document is returned to PTCC and CDER management. The final edition is published in the Federal Register as well as placed on the CDER website.

In contrast to guidances, rules are posted in the CFR and are usually negotiated with Congress. PDUFA requirements for completing IND safety reviews and formats for submitting INDs and NDAs are contained in the CFR.

The purpose of the ICH is to reach a consensus on what studies are necessary to license a drug in Europe, the United States, and Japan. The idea was that a single application package ought to be sufficient for all the regulatory agencies instead of a series of application packages with different studies for each area. The necessity of carcinogenicity studies, maximum doses for toxicity studies, duration of the longest nonrodent studies, types of genotoxicity studies, and special pharmacology studies had been included in the areas of discussion. Discussions are ongoing and guidance can change with new scientific evidence.

ICH guidances go through a few more iterations prior to publication in the Federal Register. Topics for discussion are determined jointly by the European, American, and Japanese representatives (regulatory and industry). Once the topic is decided, specific aspects of the topic are discussed. At this point, data may be gathered to support a specific position. For example, most of the European agencies required only six-month dog data, while Japan and the United States required a one-year study. In negotiating the maximum time required for chronic nonrodent toxicity studies, the U.S. FDA compiled data on toxicity findings in dogs at 6 and 12 months and determined that there were significant differences at the two time points. A compromise time of nine months was reached in the final draft of the ICH paper. The drafts from the ICH are brought to the PTCC and then back to the ICH for further discussion. This can be repeated for several cycles. Finally, the guidance goes through the CDER management and is published for comment in the Federal Register prior to the final form, which is again published in the Federal Register.

The discovery and development of new drugs is a lengthy and costly process. Pharmaceutical and biologic manufacturers must undergo a rigorous process to demonstrate the safety and effectiveness of new products, before the FDA approves the selling of such products in the United States. It should be noted however that the process is similar (if not identical) in all countries, each having their own regulatory agency or group. These regulatory agencies are essentially equivalent to the FDA. FDA goes through a meticulous review process which is a synthesis of the expertise of multiple scientific disciplines such as medicine, chemistry, biostatistics, biopharmaceutics, pharmacology, and toxicology. For physicians, this drug evaluation science is comparable to a new specialty field, and for nonphysicians it represents a similar kind of specialization within their disciplines. All evaluation and integration of the drug data are done within an established legal and administrative context. Ultimately, the logic and the conclusions reached by each evaluator scientist must be clear and compelling to the expert community.

The public's need for a better understanding of the drug evaluation process was stimulated by increasing sentiment during the early 1990s that the government has become too large, too intrusive, too expensive, and too unaccountable for its results. The Reinventing Government (REGO) movement is a direct consequence of this thinking, and REGO has presented an additional incentive for FDA to define and explain its processes. A large part of the philosophy of the reinvention movement is to make the government accountable to its customers; however, there is some debate over whether the ultimate consumer for the FDA is the citizen who uses the products regulated by the agency or the drug companies which are regulated by the agency. The PDUFA of 1992 and its recent renewal are in part responses to the REGO movement. An increased awareness of the urgency of time in the review process is only one of many such changes. PDUFA however has not meant an open-ended period of generous support for drug review. PDUFA provides only a window of opportunity for the agency to get the personnel, the processes, and the tools to do its job in the twenty-first century.

Recognition that faster drug review would involve much attention to both process and personnel is an important impetus for the FDA's involvement in drug regulatory science and the closely related concept of Good Review Practices (GRP). The FDA's achievement of high performance (speed and quality of review) in drug evaluation will require that the principles of drug evaluation are defined, understood, and practiced by agency professionals and are widely accepted and understood by those submitting data to the agency. FDA has for decades required all drug developers to follow GMP, GLP, and GCP in the manufacture and investigation of their drug. Now the agency is asking that its professionals similarly adhere to a good practice standard.

Ultimately, the goal of drug development is to obtain data needed to secure marketing approval in an orderly and logical progression. The drug evaluation science as a discipline helps the agency respond to the demand that government be more accountable for its results, not just outputs. The science-based review allows the agency to respond or advise the sponsors in a rational, informative, and knowledge driven direction. The distinction between outputs (reports filed, and reviews written) and results (disease prevented, and deaths averted) was put into law with the bipartisan support and passage of the Government Performance and Results Act in 1992. Very simply, in the context of drug review, examples of agency outputs include counts of reviews produced, meetings held, and decisions made without any evaluation of their quality or impact. These outputs may lead to benefits but, in themselves, are not actual benefits such as mortality reduction or fewer adverse drug reactions. The agency is held accountable by the public for the speed and productivity of therapeutics development almost as much as the therapeutic developers themselves. Although this expectation overestimates FDA's importance in the development of new drugs and other therapies, FDA's adjudged performance in this arena hinges again on a streamlined, efficient process understood by reviewers and developers alike. The advancement of drug evaluation science supports the expeditious development of needed therapies and other recognizable results.

Within the global free market of drug regulatory approaches, there has been international cooperation. The ICH of Technical Requirements for Licensing of Pharmaceuticals has brought together the drug regulating authorities from Europe, Japan, and the United States. The objective is to achieve a single set of requirements in the areas of manufacturing, animal testing, and clinical trials that are acceptable to all three authorities for new drug approval. Much of the discussions in ICH have been based on science, and therefore can be considered a discourse of drug regulatory science itself. ICH also has the valuable effects of highlighting the basic differences in approaches used by the three authorities and identifying those principles that can be universally utilized.

As an example, the European and the FDA approaches to new drug approval differ significantly. Europe's various national agencies and, more recently, the European Union authority, the Medicines Evaluation Agency, have a small number of review professionals who rely heavily on outside academic consultants to produce what are called expert reports. These reports are the basis of approval decisions made by committees, also composed of outside experts. In contrast, the FDA is staffed with a large number of reviewing professionals who have training and experience within their assigned areas of responsibility. While the FDA does utilize outside expertise through its advisory committees and occasionally in situations where internal expertise may require supplementation, the drug review process is performed, and decisions are made by FDA review specialists.

Probably the most fundamental difference between the European and the FDA approaches is in the depth of review. In Europe, the expert reports are generally based on data summaries. At the FDA, original source data as well as data summaries submitted by the developer are independently checked, calculated, and analyzed by agency reviewers. The FDA also has a comparatively rigorous field inspection system for the examination of animal and clinical investigation sites. The FDA, unlike Europe, is also directly involved in reviewing and authorizing early human drug studies. In Europe, these Phase I studies are generally supervised by local institutional ethics committees without direct regulatory authority involvement.

This seems more plausible when it is recognized that even though the regulatory authorities in Europe, Japan, and other countries do not perform some of the functions carried out by the FDA, these functions nonetheless are performed by other individuals, organizations, or the drug developer itself. Even with such different governmental approaches, it is certain that the evaluation scientists, whether privately or publicly supported, should be practising the same science. The appropriate application of biostatistical methodology, the use of in vitro toxicology testing, and the design and interpretation of clinical studies are examples of scientific topics that have been addressed within the ICH forum.

Review of the drug development process from the pharmacology/toxicology perspective starts with an understanding of the pharmacology of the drug; in a broad sense it deals with understanding an interaction of exogenously administered chemical molecules (drugs) with living systems. It encompasses all aspects of knowledge about the drugs, but most importantly, those that are relevant to effective and safe use for medicinal purposes. For thousands of years most drugs were crude natural products of unknown composition and having limited efficacy. Only the overt effects of these substances on the body were rather imprecisely known, but how the effect was produced was entirely unknown.

ASPECTS OF THE IND/NDA PROCESS

Pharmacology

For thousands of years compounds have been purified and characterized, and a vast variety of highly potent and selective new drugs have been developed. In the recent past, the chemical activity and the mechanism of action, including molecular targets, of many drugs have been elucidated. Pharmacology forms the backbone of rational therapeutics.

The two main divisions of pharmacology are pharmacokinetics and pharmacodynamics. Pharmacokinetics refers to the movement of the drug into the body and the alteration of the drug by the body, i.e., *what the body does to the drug*. This includes absorption, distribution, binding/localization/storage, biotransformation, and excretion of the drug. Sponsors need to provide the information regarding the absorption, distribution, metabolism, and elimination (ADME) of the compound for the proper safety evaluation as well as establishing a dose–response curve.

Pharmacodynamics is the study of drug effects, i.e., *what the drugs do to the body*, and attempts to elucidate the complete action–effect sequence and the dose–effect relationship. Modification of the effects of one drug by another drug and by other factors is also a part of pharmacodynamics. At the heart of pharmacodynamics is the dose–response relationship. Dose–response curves depict the relationship between dose of drug administered and the resulting pharmacologic effect. Logarithmic transformation of dosage is frequently used, because it permits the display of a wide range of doses. Dose–response curves are characterized by differences in (*i*) potency, (*ii*) slope, (*iii*) efficacy, and (*iv*) individual responses.

The potency of a drug is depicted by its location along the dose axis of the dose–response curve. Factors that influence the potency of a drug include (*i*) absorption, (*ii*) distribution, (*iii*) metabolism, (*iv*) excretion, and (*v*) affinity for the receptor. For clinical purposes, the potency of a drug makes little difference as long as the effective dose (ED) of the drug can be administered conveniently. Of far greater importance is the therapeutic index (TI), or ratio between the ED and the toxic dose. The dose required to produce a specified effect is designated as the ED necessary to produce that effect in a given percentage of patients (ED50, ED90). Increased affinity of a drug for its receptor moves the dose–response curve to the left.

The slope of the dose–response curve is influenced by the number of receptors that must be occupied before a drug effect occurs. For example, if a drug must occupy a majority of receptors before an effect occurs, the slope of the dose–response curve will be steep. A steep dose–response curve is a characteristic of neuromuscular blocking drugs and inhaled anesthetics (minimal alveolar concentration); it means those small increases in dose evoke intense increases in drug effect (Fig. 1).

The maximal effect of a drug reflects its intrinsic activity or efficacy. This efficacy is depicted by the plateau in dose–response curves. It must be recognized that undesirable effects (side effects) of a drug may limit dosage to below the concentration associated with its maximal desirable effect. Differences in efficacy are emphasized by the pharmacologic effects of opioids versus aspirin in relieving pain. Opioids relieve pain of high intensity, whereas maximal doses of aspirin are effective only against mild discomfort. The efficacy and the potency of a drug are not necessarily related. Thus, the TI, or margin of safety, is the difference between the dose of drug that produces a desired effect and the dose that produces undesirable effects.

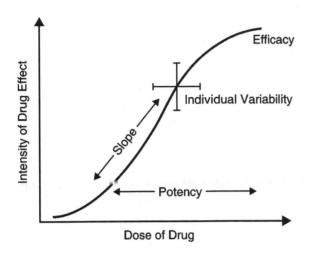

Figure 1 Dose–response curves are characterized by differences in potency, slope, efficacy, and individual responses.

In laboratory studies, the TI is often defined as the ratio between the median lethal dose and the median ED (LD50 or ED50). Drugs have multiple therapeutic indices, depending on the therapeutic response under consideration and the dose of drug necessary to evoke that response. For example, the TI for aspirin to relieve headache is greatly different from the TI to relieve the pain caused by rheumatoid arthritis.

Individual responses to a drug may vary as reflections of differences in pharmacokinetics and/or pharmacodynamics among patients (1). This may account even for differences in pharmacologic effects of drugs in the same patient at different times. The relative importance of the numerous factors that contribute to variations in individual responses to drugs depends, in part, on the drug itself and its usual route of excretion. Drugs excreted primarily unchanged by the kidneys tend to exhibit smaller differences in pharmacokinetics than do drugs that are metabolized. The most important determinant of metabolic rate is genetic. The dynamic state of receptor concentrations, as influenced by diseases and other drugs, also influences the variation in drug responses observed among patients.

Drugs (except for those gene based) do not impart new functions to any system, organ, or cell; they only alter the pace of ongoing activity. The basic types of drug action can be broadly classified as given below:

1. *Stimulation*: Selective enhancement of the level of activity of specialized cells, e.g., adrenaline stimulates heart and pilocarpine stimulates salivary glands. However, excessive stimulation is often

followed by depression of that function, e.g., a high dose of picrotoxin, a central nervous system (CNS) stimulant, produces convulsions followed by coma and respiratory depression.

2. *Depression*: Selective diminution of activity of specialized cells, e.g., barbiturates depress CNS and quinidine depresses heart. Certain drugs stimulate one type of cells but depress the other, e.g., acetylcholine stimulates intestinal smooth muscle but depresses SA node in heart. Most drugs cannot be just classified as stimulants or depressants.

3. *Irritation*: This connotes a nonselective, often noxious effect and is particularly applied to less specialized cells (epithelium, connective tissue). Mild irritation may stimulate associated function, e.g., bitters increase salivary and gastric secretion and counterirritants increase blood flow to the site. But strong irritation results in inflammation, corrosion, necrosis, and morphological damage. This may result in diminution or loss of function.

4. *Replacement*: This refers to the use of natural metabolites, hormones, or their congeners in deficiency states, e.g., levodopa in Parkinsonism, insulin in diabetes mellitus, and iron in anemia.

5. *Cytotoxic action*: Selective cytotoxic action against invading parasites or cancer cells, attenuating them without significantly affecting the host cells, is utilized for cure/palliation of infections and neoplasms, e.g., penicillin, chloroquine, mebendazole, cyclophosphamide, etc.

Barring a handful of drugs whose actions can be explained on the basis of their simple physical or chemical properties, majority of drugs act in a complex manner—all elements of which are seldom known. The fundamental *mechanisms of drug action* can be distinguished into four categories.

1. A *physical property* of the drug is responsible for its action, e.g., mass of the drug [bulk laxatives (bran), protectives (dimethicone)]; adsorptive property (charcoal, kaolin); osmotic activity (magnesium sulfate, mannitol); radioactivity (^{131}I and other radioisotopes); and radio-opacity [contrast media (barium sulfate, Urografin)].

2. A *chemical property* of the drug reacts extracellularly according to simple chemical equations, e.g., antacids ($AlOH_3$ and others) neutralize gastric HCl; acidifying (NH_4Cl) and alkalinizing ($NaHCO_3$) agents react with buffers in plasma and alter pH of urine; oxidizing agents ($KMnO_4$) are germicidal and inactivate ingested alkaloids; and chelating agents (calcium disodium edetate, British Anti-Lewisite (BAL), penicillamine) sequester toxic metals.

3. Almost all biological reactions are carried out under catalytic influence of *enzymes*; hence enzymes are a very important target

of drug action. Drugs can either increase or decrease the rate of enzymatically mediated reactions. In physiological systems enzyme activities are often optimally set. Thus, *stimulation* of enzymes by drugs, that are truly foreign substances, is unusual. Enzyme stimulation is relevant to many endogenous mediators and modulators, e.g., adrenaline stimulates adenylyl cyclase, and pyridoxine acts as a cofactor and increases decarboxylase activity. Stimulation of an enzyme increases its affinity for the substrate so that rate constant (K_m) of the reaction decreases.

Apparent increase in enzyme activity can also occur by enzyme induction, i.e., synthesis of more enzyme protein. This can not be called stimulation because the K_m does not change. Many drugs induce microsomal enzymes, e.g., methicillin induces penicillinase in some bacteria. *Inhibition* of enzymes is another common mode of drug action (Fig. 2).

A. *Nonspecific inhibition*: Many chemicals and drugs are capable of denaturing proteins. They would alter the tertiary structure

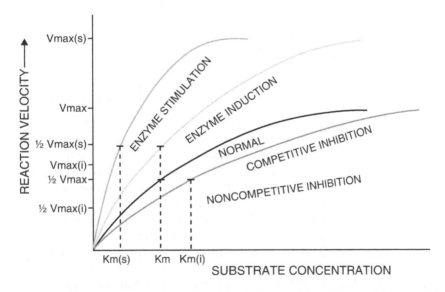

Figure 2 Effect of enzyme induction, stimulation, and inhibition on kinetics of the reaction: V_{max}—maximum velocity of reaction; $V_{max}(s)$—of stimulated enzyme; $V_{max}(i)$—in presence of noncompetitive inhibitor; K_m—rate constant of the reaction; $K_m(s)$—of stimulated enzyme; $K_m(i)$—in presence of competitive inhibitor. *Note*: Enzyme induction and noncompetitive inhibition do not change the affinity of the enzyme (K_m is unaltered), whereas enzyme stimulation and competitive inhibition, respectively, decrease and increase the K_m.

of any enzyme with which they come in contact and thus inhibit it. Heavy metal salts, strong acids and alkalis, alcohol, formaldehyde, and phenol inhibit enzymes nonspecifically.

B. *Specific inhibition*: Many drugs inhibit a particular enzyme without affecting others. Such inhibition is either competitive or noncompetitive.

   i. *Competitive* (equilibrium type): The drug competes with the normal substrate or the coenzyme so that a new equilibrium is achieved in the presence of the drug. Such inhibitors increase the K_m but the V_{max} remains unchanged, i.e., a higher concentration of the substrate is required to achieve 14 maximal reaction velocity, but if the substrate concentration is sufficiently increased, it can displace the drug and the same maximal reaction velocity can be attained.

- Physostigmine and neostigmine compete with acetylcholine for cholinesterase.
- Sulfonamides compete with PABA for bacterial folate synthetase.
- Allopurinol competes with hypoxanthine for xanthine oxidase.
- Carbidopa and methyldopa compete with levodopa for dopa decarboxylase.
- A drug may also compete with a coenzyme, e.g., warfarin competes with vitamin K which acts as a coenzyme for enzymes synthesizing clotting factors in the liver.

A nonequilibrium type of enzyme inhibition can also occur with drugs which react with the same catalytic site of the enzyme, but either form strong covalent bonds or have such high affinity for the enzyme that the normal substrate is not able to displace the inhibitor, e.g., organophosphates react covalently with the esteric site of the enzyme cholinesterase. Methotrexate has 50,000 times higher affinity for dihydrofolate reductase than the normal substrate dihydroxy folic acid (DHFA). In these situations K_m is increased and V_{max} is reduced.

   ii. *Noncompetitive*: The inhibitor reacts with an adjacent site and not with the catalytic site, but alters the enzyme in such a way that it loses its catalytic property. Thus K_m is unchanged but V_{max} is reduced. Examples are acetazolamide (carbonic anhydrase); aspirin, indomethacin (cyclo-oxygenase); nialamide, tranylcypromine (monoamine oxidase); disulfiram (aldehyde dehydrogenase); digoxin (Na^+K^+ ATPase);

theophylline (phosphodiesterase); and propylthiouracil (peroxidase in thyroid).

4. *Receptors*: A large number of drugs act through specific macromolecular components of the cells which regulate critical functions like enzyme activity, permeability, structural features, template, etc. These macromolecules or the sites on them, which bind and interact with the drug, are called "receptors."

   Receptor is defined as a binding site with functional correlate(s). Receptors are situated on the surface or inside the effector cell, and specific agonists combine with them to initiate the characteristic response. Receptors may be an agonist (it activates a receptor to produce an effect), inverse agonist (it activates a receptor to produce an effect in the opposite direction to that of the well-recognized agonist), antagonist (it prevents the action of an agonist on a receptor or the subsequent response, but does not have any effect of its own), or a partial agonist (it activates a receptor to produce submaximal effect but antagonizes the action of a full agonist). Ligand is a molecule, which attaches selectively to particular receptors or sites. The term ligand only indicates affinity irrespective of functional change: Both agonists and competitive antagonists are ligands of the same receptor.

The specificity of a compound is determined in vitro by screening the compound through a number of receptor binding or enzyme assays or more recently through a series of transcription-based assays and potential ion channel assays. However, the nonspecific tissue distribution of the enzymes and receptors as well as the transcription factors and ion channels more often results in primary as well as secondary pharmacodynamics of the compound. The sponsor needs to assess the primary pharmacodynamics, i.e., the efficacy of the compound based on its mechanism of action, e.g., morphine binds to mu opioid receptors in brain, and its primary therapeutic function is producing analgesia. Therefore, the sponsors are required to assess the potency and efficacy of the compound in the in vivo animal model to demonstrate the efficacy of the compound through the desired drug delivery system. In addition to the evaluation of the efficacy or the primary pharmacodynamics, the sponsors are also required to determine the secondary pharmacodynamics of morphine because the mu opioid receptors are distributed in the peripheral tissues as well. Therefore, using the same route of administration and the drug delivery system, the sponsors are required to estimate the plausible secondary pharmacodynamics, which in case of morphine are well characterized, such as respiratory depression, blood pressure modulation, and gastrointestinal motility. The in vivo efficacy model should provide information regarding the secondary pharmacodynamics when monitored appropriately. However, for the evaluation of the secondary pharmacodynamics in depth, organ-specific in vivo safety pharmacology evaluation of the compound should be undertaken to ensure a proper therapeutic ratio.

Safety Pharmacology

The safety pharmacology core battery is designed to examine the effects of an investigational drug on vital functions. These studies are guided by the regulations described in "Safety Pharmacology Studies for Human Pharmaceuticals, ICH-S7B, 2002." These studies are typically required for all new drugs except for topical products where the pharmacology of the drug is known and systemic exposure is minimal. These studies are also not required for drugs used in the treatment of end stage cancers.

Core systems include the CNS, the cardiovascular system, and the respiratory system. The cardiovascular system is given a special place in toxicology evaluations because a single cardiovascular event (such as arrhythmia) could result in sudden death. Cardiovascular evaluations are typically conducted, not just as a part of the safety pharmacology battery, but blood pressure, heart rate, and ECG are also typically recorded in toxicology studies in dogs. In particular, if Qt changes are observed, additional studies may be indicated. In addition, if a drug belongs to a class of compounds known to produce effects on the cardiovascular system (e.g., a negative inotropic effect), then additional testing may be indicated from the outset. Cardiac output, ventricular contractility, and vascular resistance studies are sometimes conducted. Evaluations of the CNS typically consist of reports of motor activity, behavioral changes, coordination, sensory/motor reflex responses, and body temperature effects. Typically the functional operational battery is conducted, but a number of other evaluations are also appropriate. If positive signals are observed, follow-up testing can be conducted using behavioral pharmacology, learning and memory, neurochemistry, and ligand binding studies.

The respiratory system is examined by recording the respiratory rates, tidal volume, and hemoglobin oxygen saturation. Additional studies of airway resistance, pulmonary arterial pressure, and blood gases may also be necessary depending on preliminary findings.

Although the cardiovascular, central nervous, and respiratory systems constitute the core battery of safety pharmacology testing, repeat dose toxicology studies or the pharmacology of the drug may indicate additional testing. Other systems frequently examined are the renal system, the autonomic nervous system, gastrointestinal system, skeletal musculature, immune system, and endocrine system.

Qt Prolongation

Torsade de pointes is a polymorphic ventricular arrhythmia that appears on the ECG as a continuous twisting of the vector of the QRS complex around the isoelectric baseline. This phenomenon is typically associated with the prolongation of the Qt interval and can result in life-threatening cardiac arrhythmias such as ventricular fibrillation and sudden death. Drugs that result in Qt prolongation have been either withdrawn from the market (terfenidine, cisapride, grepafloxacin) or denied approval by the FDA (lidoflazine). While a number of approved drugs are known to produce Qt prolongation (because this is a part of their mechanism as antiarrhythmic drugs), excessive prolongation can lead to new arrhythmias and death.

The seriousness of this finding has led to the publication of documents outlining the studies that are to be conducted on preclinical and clinical levels. The core battery of safety pharmacology testing includes an evaluation of the effects of the drug on blood pressure, heart rate, and ECG. In vivo, in vitro, and ex vivo evaluations of repolarization and conductance effects are also encouraged. Consideration of the pharmacological class of the compound in question and an in vivo Qt assessment (typically performed on the dogs being studied for repeat dose toxicology) will generally provide a good basis for determination if there is a risk for Qt prolongation. Care should be taken to cover the time that the drug remains in the test species and to test after repeated dosing similar to the proposed clinical regimen. In addition, early testing in the clinic should also include ECG evaluations, taking into account the pharmacokinetics of the drug.

If any positive findings are discovered, further testing is indicated (see guidance). Ionic current assays using heterologous expression systems and cell lines such as Chinese hamster ovary (CHO) cells, mouse fibroblasts (Ltk cells), and human embryonic kidney cells (HEK293) expressing hERG (human ether-a-go-go-related gene) are recommended. Purkinje fiber, papillary muscle, perfused myocardium, or even isolated intact heart may be used. In vivo studies allow the investigator to see how the drug influences the ECG in the presence of all intact (neuronal and hormonal) systems. Conscious dogs equipped with Holter monitors allow long-term evaluation of drug effects on the ECG over the full spectrum of changes in drug plasma level. Again dose ranges and duration of dosing should reflect the proposed clinical situation.

Pharmacokinetics and Toxicokinetics

Changes in the scientific emphasis in regulations can sometimes be driven by personnel changes. In the 1990s, Carl Peck, whose research interests were in the field of biopharmaceutics, was the head of the CDER at the FDA. With his interest in the pharmacokinetics, ADME, and modeling, as well as advances in the state of the science, he brought about a shift in emphasis toward having a greater dependence on the pharmacokinetic (PK) models and principles for rational drug development. Earlier submission of data on drug handling by animals and humans was expected to lead to more use of this data in picking lead compounds, designing clinical trials, selecting starting doses, and determining dose escalation schemes.

Toxicokinetics were discussed by the ICH, and a guidance on the subject was published in 1995. The basic difference between pharmacokinetics and toxicokinetics is that the latter is concerned about the handling of a drug by the body at doses with defined toxicities. Toxicokinetics is defined in the guidance as "the generation of PK data, either as an integral component in the conduct of nonclinical toxicity studies or in specially designed supportive studies, in order to assess systemic exposure" (ICH-S3A, Guideline for Industry. Toxicokinetics: The Assessment of Systemic Exposure in Toxicity Studies, March 1995). Generally, toxicokinetic data are collected during the toxicity studies; higher doses with multiple dosing schedules are investigated. In addition to allowing comparisons between toxicities and exposure levels, the data may allow selection of a more appropriate species.

For example, in oncology, it has been determined that the LD10 (a lethal dose to 10% of the animals) in rodents correlates well with the maximum tolerated dose (MTD) in humans. If the area under the curve (AUC) at the LD10 in rat is known, metabolism and protein binding are similar in rats and humans, and the AUC in humans is monitored at each dose, a more rapid dose escalation scheme might be feasible (2).

While toxicokinetic data are not absolutely required for the approval of a drug, there are multiple uses for this information. In the package insert label, the pregnancy and carcinogenicity sections compare exposure levels in the animals with teratogenic or tumor findings with the exposure at the recommended human doses to give a sense of the margin of safety. Tissue levels of compounds can yield insights into whether the accumulation of drug in a specific tissue associates with toxicity over time.

TOXICOLOGY

One of the main concerns of the regulatory agencies around the world is the safety of the compound to be administered in the human body. To date, one of the most reliable ways to assess the safety of a foreign substance is to administer it in a living system and monitor the toxicity of the substance. The first nonclinical toxicology guidance was published in 1949. Since then, the toxicity studies have undergone rigorous scrutinization regarding the concordance of the toxicity of the pharmaceuticals in humans and animals. It is generally accepted that a new molecular identity has to be tested for toxicity in two species (at least one nonrodent). The dosage of the compound to be tested in the animal should ensure adequate safety margin, and the toxic potential of the compound should be fully delineated. Generally, in toxicity studies data are incorporated from physical examination, clinical signs (preliminary behavioral change, respiratory distress, etc.), serum chemistry, hematology, urinalysis, gross pathology, organ weight changes, and histopathology. Any changes in liver/kidney enzymes depicting abnormalities in metabolism or elimination of the compound are usually captured while analyzing these parameters. Gross changes in RBCs, coagulating factors, and WBCs are also informative, and when associated with tissue changes are predictive of toxicological findings. Urinalysis is also routinely performed to ensure proper balance of ions in the body and to detect changes in kidney function.

All the above-mentioned noninvasive procedures are performed so that underlying changes in tissue histopathology (if any) can be correlated. Observations are also used to find potential surrogate markers or early indicators of organ damage in clinical trials. Gross lesions are regularly studied at autopsy. Organ weights are taken, and tissues are processed for histopathological analysis.

Any toxicological findings should be thoroughly investigated by a board-certified pathologist, and the data should be analyzed with appropriate statistics.

The dose levels in definitive toxicity studies are guided by the toxicology findings in preliminary studies and the pharmacodynamic responses of the test species. The lowest dose in a toxicity study should produce no observable adverse effects. This is called the NOAEL or no adverse effect level. The highest dose should be frankly toxic to the test system or be a maximum feasible dose so that the full toxic potential of the compound is known (a worst case scenario). The middle dose lies between the low and high doses. Exposure at each dose level can be extremely useful in interpreting the results and extrapolating it to the human condition. Both male and female animals should be utilized in the main study. It is normal to estimate exposure in animals of both sexes unless some justification can be made for not doing so.

Although estimating exposure as described above may aid in the interpretation of toxicity studies, and in comparison with human exposure, a few caveats should be noted. Species differences in protein binding, tissue uptake, receptor properties, and metabolic profile should be considered. For example, it may be more appropriate for highly protein bound compounds to have exposure expressed as the free (unbound) concentrations. In addition, the pharmacological activity of metabolites, the toxicology of metabolites, and antigenicity of biotechnology products may be complicating factors. Furthermore, it should be noted that even at relatively low plasma concentrations, high levels of the administered compound and/or metabolites may occur in specific organs or tissues.

The toxicokinetic strategy to be adopted for the use of alternative routes of administration, e.g., by inhalation, topical or parenteral delivery, should be based on the PK properties of the substance administered by the intended route. It sometimes happens that a proposal is made to adopt a new clinical route of administration for a pharmaceutical product; e.g., a product initially developed as an oral formulation may subsequently be developed for intravenous administration. In this context, it will be necessary to ascertain whether changing the clinical route will significantly reduce the safety margin.

This process may include a comparison of the systemic exposure to the compound and/or its relevant metabolites (AUC and C_{max}) in humans generated by the existing and proposed routes of administration. If the new route results in increased AUC and/or C_{max} or a change in metabolic profile, the continuing assurance of safety from animal toxicology and kinetics should be reconsidered. If exposure is not substantially greater, or different, by the proposed new route compared to that for the existing route(s), then additional nonclinical toxicity studies might focus on local toxicity.

One of the concerns for regulators is determining when the amount of animal data is sufficient to allay concerns about the safety of the compound in humans. A series of ICH guidances have been agreed upon (i) to relate the timing of the nonclinical studies to the phase of clinical development (ICH-M3) and (ii) to relate the longest duration of the study to the use of the drug (ICH-S4A). Basically, the duration of the animal study must exceed that of the proposed human trial prior to initiation of that trial. So, if a company plans to conduct a three-week trial of their new drug, they would most likely have a four-week animal study prior to starting of that human trial. The maximum duration for a study in rodents for a general toxicity study is six months. It is also expected that for most long-term use, a two-year rodent carcinogenicity study will also be conducted.

The maximum duration in a nonrodent species is currently set at nine months. This was a compromise from the ICH negotiations. Most European agencies required only six-month duration in a nonrodent study. The U.S. and the Japanese programs preferred a one-year study. The FDA looked at their files to determine if new toxicities appeared between 6 and 12 months and found that in several cases, there were significant new toxicities evident after six months (2). Depending on the risk–benefit ratio, knowledge of the drug class, and patient population, the duration of the nonrodent toxicity study may range from six months to one year. The issue is currently under re-examination with data from drugs approved since the early 1990s.

Based on the indication that the drug is intended for treatment, the starting dose determined from the toxicity studies may differ. For normal volunteers, a common approach has been delineated in the paper: Estimating the safe starting dose in clinical trial for therapeutics in adult healthy volunteers (July 2005). Here the approach is to use a fraction of the NOAEL dose from the animal studies to determine a safe first dose in humans. In oncology, where the risk–benefit ratio to patients (not normal volunteers) is much greater, a fraction of the LD10 in rodents (a dose lethal to 10% of the animals) is used to set the initial dose. This is based on data from

Freirich et al. (3) where the LD10 in rodents was determined to be similar to the MTD in humans.

Genetic Toxicology

The genetic toxicology assays are relatively quick and inexpensive in determining the potential of a compound to cause mutagenicity (e.g., damage to specific base pairs) or clastogenicity (e.g., damage to the chromosome as in chromosome breaks). Less frequently, the ability of a compound to impair DNA repair is assessed. The specific tests recommended for these assessments are delineated in "ICH-S2B: A standard battery for genotoxicity test of pharmaceuticals (July 1997)." The details of technical aspects of the individual assays are discussed in "ICH-S2A: Specific aspects of regulatory genotoxicity tests for pharmaceuticals (April 1996)."

The general battery consists of a bacterial mutagenicity assay (Ames test), a mammalian in vitro assay (either for clastogenicity, e.g., CHO chromosomal aberration assay, or mutagenicity, e.g., mouse lymphoma TK assay), and an in vivo assay, usually the mouse micronucleus assay. Other assays are still acceptable or may be used to further investigate mechanisms of genotoxicity, such as assays of DNA adducts or of cellular repair. The specific aspects of these tests, including definitions of positive and negative assays and dose selections, are covered in the ICH-S2A document.

Carcinogenicity

Given the expense and the long duration of the two-year rodent carcinogenicity tests, a great number of questions on their conduct and interpretation have arisen over the years. As requirements differed across Europe, the United States, and Japan, this topic was discussed extensively at ICH, and several guidances have been published. Whether a single species or a single gender within a species would be sufficient to determine the potential for carcinogenicity, was also debated. While there are frequently difficulties in determining the relevance of findings in the two-year studies to the human condition, this has been the "gold standard" since the 1960s. An extensive database on histopathologic findings in various subspecies of rats and mice has been established over the years.

One of the first of the ICH guidances, the S1A document "The Need for Long-Term Rodent Carcinogenicity Studies of Pharmaceuticals," 1996, discussed when two-year rodent carcinogenicity studies are necessary for drug licensing. An issue here involved the length of human exposure that triggers the need for a carcinogenicity assay. The Japanese position in the 1990s was that six months of continuous use in humans would necessitate two-year rodent studies, while the United States position was that three months of continuous human use would be sufficient for requiring carcinogenicity studies. Industry and the regulatory agencies determined that any drug that would be used for three months would likely be continued for six months, thus a six-month duration of human treatment would be sufficient to initiate carcinogenicity studies. Frequent intermittent exposures (such as allergic rhinitis, depression/anxiety) would also trigger the need for carcinogenicity studies. Every drug can be an exception to the rule though: if there is evidence of carcinogenicity in other compounds of the class, structural alerts, preneoplastic lesions in the general toxicity studies, or long-term tissue retention associated with local tissue reactions, studies may be necessary. On the other hand, where there are unequivocally positive results in genotoxicity studies, the rodent carcinogenicity studies need not be conducted. Similarly in cases where life expectancy in the human population to be treated is less than three years, no carcinogenicity studies would be needed. Issues of route versus systemic exposure, photocarcinogenicity, similar formulations, and endogenous peptides were also discussed.

The design of the two-year study is also crucial and sometimes controversial. The ICH-S1B document, Testing for Carcinogenicity of Pharmaceuticals, 1997, addresses some of the approaches for testing including selection of species, additional testing, and evaluation of carcinogenic response. Since the S1B document was published, alternative assays such as the transgenic (p53 and TGAC mouse) and neonatal mouse have been used to support the assessment of carcinogenic potential. Other models, with sufficient validation, may also be acceptable, but must be justified. As a single two-year study, unless there are clear indications otherwise, the rat is the preferred model.

Part of the logic for selection of the rat model as the default is based on a survey of the genetic toxicity, tumor incidence, pharmacology, and regulatory fate of drugs conducted by the members of the ICH. The interpretation of the data was that usually mouse tumors were not the sole reason for regulatory action, but were part of a weight-of-evidence approach. Second, twice as many compounds that were positive in a single species were rat-only as compared with mouse-only. This suggests that rat is more sensitive than mouse. Finally, mouse liver tends to be more susceptible to nongenotoxic chemicals via mechanisms not relevant in humans.

The choice of species for the two-year assay can be problematic. The largest issue is relevance in human use. Pharmacology, metabolites, exposure levels, and protein binding should all be similar to that seen in the human. It helps if there is a database of tumor findings from previous studies (historical database) that aids in determining if the finding of rare tumors is significant. Conversely, with the rodents, some frequent tumors (liver, thyroid) are caused by mechanisms that are mostly irrelevant in humans. Other mechanistic studies like measuring hormone levels or DNA adducting can help elucidate the relationship between findings in the rodents and in humans.

Suggestions for selection of the doses for carcinogenicity studies have also been addressed in ICH-S1C, Dose Selection for Carcinogenicity Studies of Pharmaceuticals. There was heated discussion over the logical upper dose limit. Issues here included what would be considered high enough to yield a meaningful test result (maximum feasible dose, dose-limiting toxicity, and some multiple of dose or exposure level) and yet low enough to ensure survival of the animals.

A PK approach was also provided in ICH-S1C (R) where a series of criteria must be met including lack of genotoxicity, similarity of metabolism, similar protein binding, and at least 25-fold above the human exposure at a maximum dose of 1500 mg/kg/day. This was based on a survey of prior carcinogenicity studies where recommended human dose, exposure levels in humans and animals at the highest dose, and genotoxicity status were known (2).

Drugs either showed at the human MTD a ratio of animal exposure to human exposure that was less than one (approximately one-third of the database of 35 compounds), or approximately 10 or less (two-thirds of the drugs examined). This suggests that an exposure ratio of 25 would be an adequate margin for evaluation of therapeutics.

With PDUFA 2, the FDA agreed to evaluate special protocols, including those for carcinogenicity studies within defined time intervals (45 days from receipt). To aid the reviewers in dealing with the data in a short time frame, guidance for what information to include in a special protocol request was published. The protocol is then presented to the Executive Carcinogenicity Assessment Committee, which consists of the Associate Director for Pharmacology and Toxicology, one member from the Office of Testing and Research, and two pharmacology/toxicology supervisors. The committee concurs or disagrees with the sponsor's protocol in the areas of species selection, appropriateness of design, and dose selection. These areas are not usually revisited in the final assessment of the study. Issues that are examined at the final submission of the completed study include whether the histopathology slides were peer reviewed, whether all tissues from all dose groups were examined, whether historical controls were included from that lab/strain, whether toxicokinetics were showing adequate exposure (usually in a satellite group), and whether the statistical methodology was appropriate and correctly conducted.

Reproductive Toxicology

The section of the package insert label regarding pregnancy is one of the most important uses of nonclinical data. The effects of a compound on the developing conceptus as well as the possible effects on fertility concern almost every user of prescription drugs. This information affects both the possibilities: whether a drug should be used at all (e.g., retinoids in women who might become pregnant) or whether to consider terminating a pregnancy (e.g., an expected pregnancy during chemotherapy). In attempting to make the information in the label more useful, not only the teratogenic changes (or lack thereof) are delineated, but comparison of the animal dose where the findings are observed to the expected clinical dose is also provided. The patient or doctor can then make their own informed decision on drug use.

Reproductive changes are seldom reported in early clinical studies, due largely to exclusion of women of childbearing potential from these experiments. Conventional toxicology studies carried out prior to the first human studies include histopathological examination of both the female and male reproductive tracts, incorporating examination of testes and ovaries. Additional reproductive toxicity studies, which include structural and functional alterations that may affect reproductive competence in the first generation, should be considered. The reproductive and the developmental toxicities are both considered in the reproductive toxicity studies. The three classes of reproductive toxicity include effects on fertility, parturition, and lactation. Male reproductive toxicity associated with administration of a drug may be seen as degeneration or necrosis of the reproductive organs, reduction in sperm count, alterations to sperm motility or morphology, aberrant mating behavior, altered ability to mate, alterations to endocrine function, or overall reduction in fertility. The duration of dosing for these studies has been a source of much discussion at the ICH. Initially, the males were dosed for 10 weeks prior to mating to affect several cycles of sperm production. The main focus of measurement was the mating/fertility index. However, a comparison of males who underwent two, four, and nine weeks of treatment prior to mating showed no significant differences between those treated for two and nine weeks. ICH now recommends treatment for two weeks. The major change in the recommendation is that histopathologic observations are the most sensitive and informative parameter to examine [ICH-S5A: Guideline for industry: detection of toxicity to reproduction for medicinal products (September 1994); ICH-S5B: Guideline for industry: detection of toxicity to reproduction for medicinal products: addendum on toxicity to male fertility (April 1996), further amended in November 2000]. Female reproductive toxicity may be seen as damage to the reproductive organs, alterations to endocrine regulation of gamete maturation and release, aberrant mating behavior, altered ability to mate, or overall reduction in fertility. Diminished fertility in female animals is typically detected by reductions in the fertility index, the number of implantation sites, and time to mating. Toxicities affecting labor and delivery in animals may be seen as changes in the onset or duration of parturition. Changes in the duration of parturition are frequently reported as mean time elapsed per pup or total duration of parturition. Drugs administered to lactating animals may be a source of unwanted exposure in the nursing neonate, may alter the process of lactation in the nursing mother (e.g., the quality or quantity of milk), or may alter maternal behavior toward the nursing offspring.

Developmental toxicities are generally those that affect the F1 generation. The four classes of developmental toxicity are mortality, dysmorphogenesis (structural alterations), alterations to growth, and functional toxicities. Mortality due to developmental toxicity may occur at any time from early conception to postweaning, ("embryo-fetal death" is a subset of mortality due to developmental toxicity). Thus, a positive signal may appear as pre- or peri-implantation loss, early or late resorption, abortion, stillbirth, neonatal death, or peri-weaning loss. Dysmorphogenic effects are generally seen as malformations or variations to the skeleton or soft tissues of the offspring, and are commonly referred to as structural alterations. Alterations to growth are generally seen as growth retardation, although excessive growth or early maturation may also be considered alterations to growth. Body weight is the most common measurement for assessing growth rate. Crown–rump length and ano-genital distance may also be measured. Functional toxicities could include any persistent alteration of normal physiologic or biochemical function, but typically only developmental neurobehavioral effects and reproductive function are measured. Common assessments include locomotor activity, learning and memory, reflex development, time to sexual maturation, mating behavior, and fertility.

A signal of reproductive as well as developmental toxicity should be assessed in relevance to the magnitude of adverse effects in the offspring versus the severity of maternal (and, for fertility studies, paternal) when drawing a conclusion about the importance of the Fo toxicity to effects observed in the offspring. This evaluation is relevant to all seven classes of reproductive and developmental toxicity. A positive signal occurring at doses that are not maternally toxic increases concern for human reproductive or developmental toxicity. If a positive signal is observed only in the presence of frank maternal toxicity, there is a decreased concern, provided that the positive signal may be reasonably attributed to maternal toxicity. When evaluating a positive signal in two or more species, assessment of the implications of maternal or paternal toxicity should be based on a composite analysis of the data from all adequately studied species. If a positive signal is seen in two or more species in the absence of maternal toxicity, there is an increased concern for adverse human reproductive outcomes. If a positive signal is seen only in the presence of clear relevant maternal toxicity in multiple species, there is a decreased concern. If there is nonconcordance between test species as to the presence and relevance of maternal toxicity, there may be no change in the overall level of concern for this contributory element.

Concern for human reproductive or developmental toxicity is increased when a positive signal is characterized by any of the following: (*i*) increased severity of adverse effects with an increase in dose, (*ii*) increased incidence of adverse effects with an increase in dose, or (*iii*) a high incidence of adverse effects across all dosed groups. Another major concern in the developmental toxicity studies is the occurrence of rare events. Developmental toxicity studies usually lack the statistical power to detect subtle increases in rare events (described in ICH). Thus, an increased frequency of positive signals for rare events in drug-exposed animals increases concern for reproductive or developmental toxicity in humans. The absence of an increased frequency of rare events, however, does not decrease concern.

A positive signal in reproductive toxicity studies should be analyzed with respect to the following three pharmacodynamic elements: (*i*) the therapeutic index (TI), (*ii*) biomarkers as a benchmark, and (*iii*) the similarity between the pharmacologic and toxicologic mechanisms. The TI is used to identify the extent to which there is an overlap between therapeutic doses and doses that cause reproductive or developmental toxicity. It is unusual to obtain well-defined dose–response curves for toxicity and efficacy from a single species. Thus, the use of estimations or surrogate endpoints (related to the therapeutic mechanism) for this evaluation may be warranted. There may be circumstances in which an effect on a biomarker is consistently seen in multiple species at doses lower than the NOAEL for demonstrable reproductive/developmental toxicity. If there is an effect on this biomarker at or below the therapeutic dose in humans, there is an increased concern for reproductive or developmental toxicity in humans. If this biomarker is responsive to the drug in humans, can be monitored, and is not affected at the therapeutic dose, there may be decreased concern.

Similarity between pharmacologic and reproductive developmental toxicologic mechanisms (e.g., delay of parturition by drugs known to suppress uterine smooth muscle contractility or hypotension in the offspring of dams treated during late gestation with a drug known to lower blood pressure) increases the concern for reproductive or developmental toxicity in humans. There is less concern if the positive signal is attributed to an animal-specific pharmacological response, even though it may be an extension of the pharmacological effect of the drug (e.g., pregnancy loss in rats due to hypoprolactinemia). Specific mechanistic studies are recommended to address the increase in concerns. However, well-analyzed results from a battery of testing for the male and female fertility and developmental toxicity studies in at least two species could be a good predictor of human reproductive toxicity.

OTHER ISSUES

As questions arise or needs change, guidance may be written to address frequently asked questions. In the 1990s, the botanical drug industry grew precipitously. These compounds were unregulated by the FDA until they made claims to treat a specific disease. At that point, they were subject to the rules for drug safety and efficacy. The Dietary Supplement Health Education Act removed botanical products, vitamins, and traditional medicines from FDA regulation. Guidance had been under development from the early 1990s on the development of botanical/traditional compounds that were traditionally used by humans, for licensing as drugs. Because these compounds have been in human use over a long period of time, acute toxicities have been delineated. The basic philosophy is to allow a Phase II trial to proceed and demonstrate efficacy (or not), then conduct the chronic, genotoxic, and reproductive studies.

After 9/11, a growing awareness of vulnerability to attack by biological weapons has prompted interest in developing antidotes. Testing for efficacy of anthrax or other biological agents in humans would not be ethical. Thus, the Bioterrorism Act of 2002 allows drugs to be approved for an indication based on testing of efficacy in animals only. Safety testing for some agents could still be done in humans, where possible, but the major proof of concept would be in animal infection models.

PEDIATRIC DRUG PRODUCTS

The vast majority of drugs marketed in the United States and listed in the Physicians Desk Reference have no information on safety and/or efficacy in pediatric use. Due to ethical concerns about consent, most drug testing was only conducted in adults. The assumption was that children were mostly smaller versions of the adult. Newer data have shown that small children, and infants especially, have different means of handling drugs (pharmacokinetics, metabolism) as well as different toxicities to developing organ systems. In 1997, the FDA proposed its Pediatric Rule (62 CFR 43900, August 15, 1997), which required manufacturers to assess the safety and efficacy of certain human drugs and biologics in pediatric patients. Once this rule was enacted (April 1999), any application for approval of a new active ingredient, new indication or biologic or new route of administration was expected to contain studies to assess the safety and effectiveness of the drug or biologic in pediatric patients. In 1997, Congress passed the FDA Modernization Act of 1997, which, among other things, authorized specific market exclusivity incentives to manufacturers who conducted pediatric studies of their drugs and submitted these to the FDA and who met certain statutory criteria. These exclusivity incentive provisions of FDAMA expired on January 1, 2002.

On January 4, 2002, a new law, the Best Pharmaceuticals for Children's Act (BPCA) was signed. This law essentially reauthorized the exclusivity incentives provided under FDAMA and created an additional mechanism for obtaining information on the safety and effectiveness of drugs when used in the pediatric population. The BPCA authorized several NIH funding mechanisms to fund studies of certain drugs if the manufacturers refused to conduct such studies. These funding mechanisms do not extend to biologics or certain antibiotics. BPCA is set to expire in 2007. On October 17, 2002, a U.S. District Court struck down the 1998 Pediatric

Rule. On December 3, 2003, the Pediatric Research Equity Act (PREA) was signed into law. This law was broader than BPCA in that it affected both drugs and biologics. Under PREA, pediatric assessment was required for certain applications unless waived or deferred. The submission is also to contain data supporting dosing and administration in each subpopulation.

Assessment of the safety of drugs for use in pediatric patients also includes preclinical testing. In contrast to clinical testing, these studies will allow a company to get an idea about the potential toxic effects after high doses of a drug in pediatric patients and also allows the examination of any histopathological changes that are associated with drug use. The FDA is currently developing guidelines for the conduct of toxicology studies in juvenile animals.

DRUG COMBINATIONS

The excitement over azidothymidine (AZT) was palpable. In patients infected with HIV, this formerly discarded now-revived anticancer drug had just been shown to reduce progression to AIDS and death. While no one dared to think out loud that this would be the cure for AIDS, everyone was hoping, and for a while it seemed like this could be it. Soon enough, it began to be clear that while AZT made a striking difference in the short term, the effects were only short lived. While other drugs were being developed, they too seemed to have limited efficacy over time. The virus was becoming resistant to the effects of the drug. Combination therapy seemed to provide the solution. These would attack the AIDS virus at multiple targets and so hopefully stall if not eliminate the virus. In 1997 the FDA approved the first antiviral drug combination, Combivir, which is a combination of zidovudine and lamivudine. In 2004, the use of monotherapy for HIV seemed like ancient history. Current medical practice consistently uses combination therapy to treat HIV/AIDS, with three, four, and even five drug therapies being used. But as the drug combinations constituted a new weapon to attack against HIV disease, the combination also meant a new set of adverse events.

The FDA has developed regulations for the preclinical assessment of the toxicity of combination products. These proposed studies will help to determine whether a drug in the combination affects the other's metabolism and if the toxicity of the combination is an additive or synergistic. The ICH guidance "The safety evaluation of drug combinations" discusses the evaluations necessary to assess these combinations.

EXCIPIENTS AND REFORMULATIONS

As technology improves and better chemicals are developed, products are often reformulated to produce drugs which cure more quickly or relieve symptoms faster or deliver drugs directly to the site of infection. For example, conventional amphotericin has been reformulated into liposomal amphotericin, which produces an altered PK profile and delivers more drug to the site of infection. Intravaginal suppositories for fungal infections may be formulated as creams to relieve vulvar itching. AIDS drugs such as saquinavir are reformulated to produce greater bioavailability when administered alone. The "Guidance for the evaluation of reformulated products" and "Nonclinical studies for the development of Pharmaceutical excipients" provide a discussion of the relevant studies. Typical recommendations include assessment of the compatibility with blood for intravenous reformulations, photoirritation studies if the reformulation absorbs UV light, and photocarcinogenicity if the product is to be used extensively on sun-exposed skin.

While the final NDA reviews are available through the Freedom of Information laws, the part of the review process that the public and the health practitioners see is the package insert label. The labels are also the text published in the Physician's Desk Reference. The content is mandated in the CFR. For the pharmacologist/toxicologist, the major sections required are "Carcinogenesis, Mutagenesis, and Fertility" and "Pregnancy" and "Nursing" sections. Here, the data obtained in the genotoxicity studies (which tests, results), carcinogenicity studies (if conducted), and reproductive toxicity studies are reported. The doses where findings are made are related to the human dose. For example, the teratogenic findings in rabbit are listed, and the dose at which they are found are compared to the recommended human dose (e.g., at X-fold the human dose on a body surface area basis). This allows the reader to get a sense of the margin of safety for a given result.

Individual divisions may also have specific guidances or points to consider for their drug classes. Delivery systems, routes of administration, and starting doses may be discussed.

CONCLUSIONS

The primary concern in the approval of drugs by the FDA, as mandated by Congress, is the safety and efficacy of the compound in the intended population. Historically, regulatory requirements have been determined by the reaction to a public health crisis. A major part of ensuring safety is using animal models to predict the toxic effects in humans. A standard battery of nonclinical studies has been delineated through a combination of science and negotiation with other international regulatory agencies. As the knowledge base changes (and present policies are re-evaluated with emerging data), requirements change to reflect the state of the science. How the work is done may sometimes reflect the political climate, but the driving force should always be maintaining the safety of the citizenry through application of state-of-the-art knowledge.

REFERENCES

1. Collins JM, Grieshaber CK, Chabner BA. Pharmacologically guided Phase I clinical trials based upon preclinical drug development. J Natl Cancer Inst 1990; 82:1321.
2. Contrera JF, Jacobs AC, Hullahalli RP, Mehta M, Schmidt WJ, DeGeorge JA. Systemic exposure based alternative to the maximum tolerated dose for carcinogenicity studies of human therapeutics. J Am College Toxicol 1995; 14:1–10.
3. Freirich EJ, Gehan EA, Rall DP, Schmidt LH, Skipper HE. Quantitative comparison of toxicity of anticancer agents in mouse, rat, hamster, dog, monkey and man. Cancer Chemother Rep 1966; 50:219–244.

6

National Toxicology Program

Mary S. Wolfe and Christopher J. Portier

National Institute of Environmental Health Sciences, National Institutes of Health, Research Triangle Park, North Carolina, U.S.A.

INTRODUCTION

More than 80,000 chemicals are registered for use in the United States and the impact of many of these chemicals on public health is unknown (1). However, safeguarding public health depends on identifying the potential toxic effects of these chemicals and the levels of exposure at which they may become hazardous to humans. In the late 1970s, interest arose within the Department of Health, Education, and Welfare for the creation of a centralized and coordinated effort for the research and testing of chemicals of public health concern. In response, Joseph A. Califano, Jr., Secretary of Health, Education, and Welfare [known today as the Department of Health and Human Services (DHHS)] established the National Toxicology Program (NTP) within the Public Health Service in 1978. NTP was created as a cooperative effort to improve coordination and integration of toxicology testing activities across the federal government, provide needed information to regulatory and research agencies, develop and validate improved testing methods, and strengthen the science base in toxicology. David P. Rall, who was Director of the National Institute of Environmental Health Sciences (NIEHS) of the National Institutes of Health (NIH), was designated to also serve as the first NTP Director (2). Secretary Richard S. Schweiker granted permanent status to the NTP in October 1981 (3). The need for a program like the NTP arose because of increasing scientific, regulatory, and congressional concerns in the 1960s to 1970s about the human health effects of chemical agents in our environment. Many human diseases were thought to be directly or indirectly related to chemical exposures; therefore, it was thought that decreasing or eliminating human exposures to those chemicals would help prevent some human diseases and disabilities (4).

For more than 25 years, the NTP has become a focal point within the U.S. government for evaluating the potential hazard of chemical and physical agents that we encounter in our daily lives and our environment. NTP's mission is to evaluate agents of public health concern by developing and applying tools of modern toxicology and molecular biology (5). Its goals are to coordinate toxicological testing within DHHS, strengthen the science base in toxicology, develop and validate improved testing methods, and provide information about potentially hazardous chemicals to health regulatory and research agencies (6). Central priorities of the program have been to obtain the best science using the best research strategies and technologies and to maintain all activities open to public scrutiny, including communication with all interested parties. NTP has drawn strength and direction from the commitment of its scientists to exchange information openly, maintain impartiality, and apply rigorous scientific peer review (1,7).

NTP plays a critical role in responding to toxicological concerns pertinent to pubic health and in providing information about alternative methods for toxicity screening and, as such, has an important, although indirect role, in shaping public health policy (1,8). This chapter will highlight some examples where programmatic activities influence the scientific decision making of national and international groups.

ORGANIZATION AND OVERSIGHT

NIEHS of NIH is the administrative headquarters of the NTP, which comprises the relevant components of three agencies—the NIEHS, the National Institute for Occupational Safety and Health (NIOSH) of the Centers for Disease Control and Prevention (CDC), and the Food and Drug Administration's (FDA) National Center for Toxicological Research (NCTR). The National Cancer Institute (NCI) of the NIH was a charter agency of the NTP and continues to serve on the NTP Executive Committee (1). NTP's activities, supported through voluntary allocations of staff, space, and resources by the participating agencies, are planned and carried out as a coordinated whole by the program director who also serves as director of the NIEHS (4).

NTP relies upon several advisory groups to provide input to the programs on its list of activities. NTP Executive Committee[a] provides oversight

---

[a] The voting members of the NTP Executive Committee consists of the heads or their designees from the Agency for Toxic Substances and Disease Registry of the CDC, CPSC, EPA, FDA, National Center for Environmental Health of the CDC, NCI of the NIH, NIEHS of the NIH, NIOSH of the CDC, OSHA of the Department of Labor.

to the NTP on policy issues. This committee is composed of the heads of federal research and regulatory agencies (or their designees) (2) and has expanded its membership since the program was originally established. The Environmental Protection Agency (EPA), Occupational Safety and Health Administration (OSHA), and the Consumer Product Safety Commission (CPSC) were invited to join so that regulatory concerns could be better communicated within the program. The committee brings together the Department's expertise in toxicology, and the regulatory agencies' experience and concern for a coordinated approach to toxicology research and testing of chemicals that will provide the information needed for controlling exposure to hazardous substances and protecting public health (4). NTP also relies upon two external federally chartered advisory groups for advice. NTP Board of Scientific Counselors and the Scientific Advisory Committee on Alternative Toxicological Methods assure regular scientific and public peer review and input on NTP activities and priorities (7).

ROLE IN SHAPING PUBLIC POLICY

At the national level, the NTP has developed an increasingly interactive relationship with governmental agencies that rely upon data from the NTP to make credible decisions that protect public health (6) without increasing the regulatory burden on the U.S. industry (6). NTP maintains a balanced research and testing program to provide data needed on a wide variety of issues that are important to public health (8). The program strives to generate the types of data that are consistent with regulatory agencies' guidelines and are relevant to their needs, as well as data that will provide insights into how a chemical or physical agent produces its effect. NTP also provides guidance on the appropriate translation of toxicological, epidemiological, and basic research data into public health decisions (1). NTP plays an important role in fostering interagency collaborations in research and exposure assessment, providing information to regulatory agencies about alternative methods for toxicity screening, and exploring new technologies for evaluating how environmental agents cause disease (8).

NTP coordinates workshops and conferences that provide opportunities to bring together researchers, regulators, policy makers, and the public to discuss issues, exchange information, or reach agreements about areas of toxicology that might potentially impact public health decisions (9). Some conference topics in the recent past include the assessment of health effects from exposure to methylmercury (November 1998) (10), state of knowledge about potential toxicity of medicinal herbs and research needs (September 1998) (11,12), the role of human exposure assessment in the evaluation of risks from environmental exposures (September 1999) (13), low-dose effects of endocrine-disrupting agents on reproduction and development (October 2000) (14,15), assessment of the allergenic potential of genetically modified foods (December 2001) (16), the utility of genetically modified mouse (GMM) models for cancer hazard identification (February 2003) (17), and the role of thyroid hormones in reproductive health (April 2003) (18). Each of these meetings provided a forum to openly exchange information and debate issues that can potentially impact human health and the environment.

At the request of the EPA, the NTP convened a scientific panel to review the scientific evidence for reported low-dose reproductive and developmental effects and dose–response relationships of endocrine-disrupting chemicals in mammalian species to assess effects on human health. EPA intended to use this information to help guide the agency in choosing appropriate assays for its Endocrine Disruptor Screening Program and in developing standardized test protocols. This meeting fostered open exchange and independent reanalysis of data from selected studies, and provided an opportunity for public input. NTP transmitted the panel's final report to the EPA in August 2001 (15,19).

One of the NTP's goals is to develop and validate new testing methods that will improve the ability to screen environmental agents for their potential toxicity and/or provide more information about how the agents cause disease. NTP invested considerable time and resources in addressing whether results from studies conducted in GMM or "transgenic" models are useful for identifying chemicals presumed to be of carcinogenic risk to humans. The goal was to determine whether GMM models might be integrated into NTP research and testing activities (17,20). During 2002 and 2003, the NTP began a formal evaluation of the data from studies conducted in GMM models for carcinogen identification, including the interpretation and reporting of results. At the onset, NTP staff examined the sensitivity of GMM models versus traditional rodent models used in the NTP's two-year toxicology and carcinogenicity studies for identification of human carcinogens recognized by the International Agency for Research on Cancer (IARC) and the Report on Carcinogens (RoC) (21). NTP also had discussions with its Executive Committee and its Board of Scientific Counselors, and held a workshop in February 2003 that brought together members of NTP advisory committees, staff from NTP-participating regulatory and science agencies, and representatives from animal welfare groups, foreign governments, and the pharmaceutical, chemical, and academic communities. NTP's intent was to bring all interested parties together (*i*) to debate the scientific interpretation of results and the appropriate mechanism for reporting findings from studies in GMM models and (*ii*) to engage the regulatory community because changes in NTP testing strategies to include GMM models would impact the science base available for cancer identification and risk estimation (17). Outcomes of this review included the initiation of a new NTP Technical Report series for genetically modified models and an improved understanding about how to appropriately apply the NTP's categorical system for levels of carcinogenic activity (22,23) to the findings from studies in GMM models.

NTP interacts with other national agencies and international organizations to develop scientifically based hazard evaluations that are accepted by other countries, and participates in the preparation of national and international toxicity testing guidelines (8). The NTP is increasingly active in developing international partnerships to establish efficient means for avoiding duplication of effort in toxicology testing, and for harmonizing method and testing portfolios. This includes collaborations with the World Health Organization (WHO), the European Ramazzini Foundation of Oncology and Environmental Sciences, and the Korea NTP (9).

NTP TOXICOLOGY AND CARCINOGENICITY TESTING PROGRAM

NTP has a broad mandate to provide toxicological characterizations for chemicals and physical agents of public health concern, and continually solicits and reviews nominations for study. Nominations undergo several levels of review with an opportunity for public comment before the NTP selects agents for study and designs and implements toxicology studies. This process helps ensure that the NTP's testing program addresses toxicological concerns pertinent to public health, and maintains balance among the types of substances evaluated (7). NTP selects agents for study based on specific criteria (*i*) testing of hypotheses that will enhance the predictive ability of future NTP studies, (*ii*) concern for widespread human exposure, (*iii*) inadequate existing scientific data, (*iv*) need for information about structurally related chemicals likely to affect human health, (*v*) amount produced or used, (*vi*) significant chemical or physical properties, (*vii*) interest by regulatory or research organizations, and (*viii*) possible public or occupational health significance (8).

The bioassay program, initiated by the National Cancer Institute (NCI) of the NIH, was transferred to the NIEHS in July 1981. This combined the chemical carcinogenicity testing efforts from NCI with the NIEHS toxicity testing and methods development activities under the NTP (4). NTP has a sound reputation in the design, conduct, interpretation, and reporting of toxicological studies, and its two-year bioassay is recognized both nationally and internationally for cancer identification (6,23,24). The objective of these two-year studies is to identify chemicals that may be carcinogenic to humans (8). NTP also conducts studies on substances of public health concern to evaluate a variety of other health-related effects, such as general toxicity, reproductive and developmental toxicity, genotoxicity, immunotoxicity, and neurotoxicity (9).

In general, the chemicals selected for carcinogenicity evaluation are studied in a sequence of subacute (14-day exposure), prechronic (3-month exposure), and chronic (2-year exposure) exposure studies. For each agent studied, a project leader designs a comprehensive testing strategy to address the identified research and testing needs (8,25). Details of the protocols and design considerations are published (25) and outlines of study protocols are available on the NTP web site along with data from NTP studies (26). Considering the large number of chemicals in use, the NTP continually sets priorities and develops strategies for toxicological characterization and hazard identification that will provide additional or better information on which regulatory decisions are based. The two-year bioassay has been strengthened by the addition of technologies and strategies that provide insights into the molecular and biological events associated with the agent's toxic effect(s) and mechanistic information useful to regulatory agencies for understanding the relevance of these effects in animals to humans (8,27).

NTP reports the findings from its long-term toxicology and carcinogenicity studies in the NTP Technical Report series and notes the strength of the evidence. NTP's Board of Scientific Counselors Technical Reports Review Subcommittee, a standing subcommittee, peer reviews the findings and conclusions in open public meetings. When necessary, the subcommittee is supplemented with ad hoc experts to encompass the appropriate expertise for review of specific studies. Five categories of evidence of carcinogenic activity are used to summarize the strength of the experimental evidence observed in an individual study of a chemical or physical agent. These categories of interpretive conclusions were first adopted in June 1983, revised in March 1986 (23,28,29), and are defined on the NTP web site (30). In 1987, this category system was applied to classify the conclusions from earlier studies (Technical Reports No. 2–200 and 202–205) (22).

NTP has published more than 300 reports of the findings and conclusions from its toxicology and carcinogenicity studies. Combined with the 200 studies conducted under the NCI testing program (31), this technical report series encompasses over 500 studies. In addition, approximately 60 reports have been published from the short-term toxicity studies in a separate series. Both these series are included in the National Library of Medicine's PubMed Database. In addition to publication in NTP Technical Reports, findings from NTP studies are also published in scientific journals. As an outcome of the review of transgenic models, the NTP began a new technical report series in May 2003 that contains the toxicology and carcinogenicity studies conducted in genetically modified models (32). Abstracts and full reports, as completed, are electronically available free-of-charge on the NTP web site (26) and from *Environmental Health Perspectives*, the NIEHS scientific journal (33).

Use of Findings from NTP Toxicology and Carcinogenicity Studies by the IARC

IARC, part of the WHO (34), coordinates and conducts research on the causes of human cancer and the mechanisms of carcinogenesis, and develops scientific strategies for cancer control. In 1969, the IARC initiated a program on the evaluation of the carcinogenic risk of chemicals to humans that later expanded to include other types of exposures, such as mixtures and physical and biological agents. IARC convenes international working groups, which have included NTP scientists in carcinogenesis and related fields, to evaluate the data on carcinogenicity and publishes as monographs the results of those critical reviews and evaluations of evidence on the carcinogenicity of a wide range of agents to which humans are or may be exposed. *IARC Monographs* are invaluable sources of information both for researchers and for national and international authorities (35,36). In the first 82 volumes of the Monograph series, some 885 agents (chemicals, groups of chemicals, complex mixtures, occupational exposures, cultural habits, and biological or physical agents) have been evaluated; a list of the agents and their categorization for carcinogenic hazard is available on the IARC web site (37).

Because IARC decisions regarding the carcinogencity of agents impact U.S. regulatory decisions and state legislation, one might wonder to what extent NTP studies have been included in IARC evaluations. Of the 885 agents evaluated, 180 NTP chemicals were reviewed citing 218 NTP technical reports (Table 1). The majority of the chemicals (118) were reported in the NTP Technical Report series on chronic toxicology and carcinogenicity studies. In a few incidences, the NTP has studied a chemical evaluated by IARC, but the report was not available at the time of the IARC review.

Use of NTP Toxicology and Carcinogenicity Studies by Regulatory Agencies

One of the goals of the NTP is to provide data to regulatory agencies for their use in making credible decisions regarding public health. The actual impact of NTP studies on regulatory decisions has not been quantified; however, this section provides some examples of where findings from NTP toxicology and carcinogenicity studies are being used.

U.S. Environmental Protection Agency

EPA uses information from the NTP in making decisions that protect human health and the environment. For example, the NTP is currently coordinating an initiative to provide the EPA with data useful in setting standards for finished drinking water. The EPA Office of Research and Development's National Center for Environmental Assessments prepares and maintains the electronic database Integrated Risk Information System (IRIS) accessible from the Internet (38). This database contains information on human health effects that may result from exposure to various chemicals in our environment (39). IRIS contains the agency's consensus scientific

Table 1   Agents for Which NTP Studies Were Included in Evaluations by IARC[a]

Aldicarb (116-06-3)	1,3-Dichloropropene (542-75-6)	Nitrofurantoin (67-20-9)
Aldrin (309-00-2)	Dichlorvos (62-73-7)	Nitromethane (75-52-5)
Allyl chloride (107-05-1)	Dicofol (115-32-2)	1-Nitronaphthalene (86-57-7)
Allyl isothiocyanate (57-06-7)	Dieldrin (60-57-1)	N-nitrosodiphenylamine (86-30-6)
Allyl isovalerate (2835-39-4)	Diethanolamine (111-42-2)	o-Nitrotoluene (88-72-2)
2-Aminoanthraquinone (117-79-3)	Di(2-ethylhexyl)adipate (103-23-1)	5-Nitro-o-toluidine (99-55-8)
1-Amino-2-methylanthraquinone (82-28-0)	Di(2-ethylhexyl)phthalate (117-81-7)	Ochratoxin A (303-47-9)
2-Amino-4-nitrophenol (99-57-0)	N,N'-Diethylthiourea (105-55-5)	Oxazepam (604-75-1)
2-Amino-5-nitrophenol (121-88-0)	Diglycidyl resorcinol ether (101-90-6)	Paracetamol (acetaminophen) (103-90-2)
2-Amino-5-nitrothiazole (121-66-4)	3,3-Dimethoxybenzidine-4,4-diisocyanate (91-93-0)	Parathion (56-38-2)
11-Aminoundecanoic acid (2432-99-7)	N,N-dimethylaniline (121-69-7)	Pentachloroethane (76-01-7)
Azacitidine (320-67-2)	2,6-Dimethylaniline (2,6-xylidine) (87-62-7)	Phenazopyridine hydrochloride (63-92-3)
Benzene (71-43-2)	Dimethyl hydrogen phosphite (868-85-9)[b]	Phenolphthalein (77-09-8)
Benzofuran (271-89-6)	3,5-Dinitrotoluene (618-85-9)	Phenoxybenzamine hydrochloride (63-92-3)
p-Benzoquinone dioxime (105-11-3)	Disperse blue 1 (2475-45-8)	o-Phenylphenol (90-43-7)
Benzyl acetate (140-11-4)	Disperse yellow 3 (2832-40-8)	Phenytoin (57-41-0)
2,2-bis(Bromomethyl)propane-1,3-diol (3296-90-0)	1,2-Epoxybutane (106-88-7)	Picloram (1918-02-1)
bis(2-Chloro-1-methylethyl)ether (108-60-1)	Ethyl acrylate (140-88-5)	Piperonyl butoxide (51-03-6)
Bromodichloromethane (75-27-4)	Ethylbenzene (100-41-4)	Polybrominated biphenyls (59536-65-1)

Bromoethane (74-96-4)

Bromoform (75-25-2)
1,3-Butadiene (106-99-0)
Butylated hydroxytoluene (128-37-0)
γ-Butyrolactone (96-48-0)
Caprolactam (105-60-2)
Captan (133-06-2)
Chlordane (57-74-9)
Chlorendic acid (115-28-6)
p-Chloroaniline (106-47-8)
Chlorobenzilate (510-15-6)
Chlorodibromomethane (124-48-1)

Chloroethane (75-00-3)
2-Chloronitrobenzene (88-73-3)
4-Chloronitrobenzene (100-00-5)
1-Chloro-2-methylpropene (513-37-1)
3-Chloro-2-methylpropene (563-47-3)
4-Chloro-m-phenylenediamine (5131-60-2)
4-Chloro-o-phenylenediamine (95-83-0)
Chlorothalonil (1897-45-6)
5-Chloro-o-toluidine (95-79-4)
CI acid orange (6373-74-6)
CI acid red 114 (6459-94-5)

Ethylene dibromide (106-93-4)

Ethylene oxide (75-21-8)
Ethylenethiourea (96-45-7)
Eugenol (97-53-0)
Fluometuron (2164-17-2)
Fumonisin B1 (116355-83-0)
Furan (110-00-9)
Furfural (98-01-1)
Furosemide (54-31-9)
HC blue no. 1 (2784-94-3)
HC blue no. 2 (33229-34-4)
HC red no. 3 (2871-01-4)

HC yellow no. 4 (59820-43-8)
Heptachlor (76-44-8)
Hexachlorobutadiene (87-68-3)
Hexachloroethane (67-72-1)
Hexachlorophene (70-30-4)
Hydrochlorothiazide (58-93-5)
Hydroquinone (123-31-9)
Isophosphamide (3778-73-2)
Isoprene (78-79-5)
d-Limonene (5989-27-5)
Magenta (631-99-5)

Procarbazine hydrochloride (366-70-1)
Propylene (115-07-1)
Propylene oxide (75-56-9)
Pyridine (110-86-1)
Quercetin (117-39-5)
Reserpine (50-55-5)
Resorcinol (108-46-3)
Riddelline (23246-96-0)
Styrene (100-42-5)
Sulfafurazole (127-69-5)
Sulfallate (96-06-7)
2,3,7,8-Tetrachlorodibenzo-p-dioxin (1746-01-6)

1,1,1,2-Tetrachloroethane (630-20-6)
1,1,2,2-Tetrachloroethane (79-34-5)
Tetrachloroethylene (127-18-4)
Tetrachlorvinphos (961-11-5)
Tetrafluorethylene (116-14-3)
Tetranitromethane (509-14-8)
4,4'-Thiodianiline (139-65-1)
Thiotepa (52-24-4)
Titanium dioxide (13463-67-7)
Toluene (108-88-3)
Toluene diisocyanates (mixture of 2,4- and 2,6-toluene diisocyanate) (26471-62-5)

(Continued)

Table 1 Agents for Which NTP Studies Were Included in Evaluations by IARC[a] (*Continued*)

CI basic red 9 (569-61-9)	Malathion (121-75-5)	Toxaphene (8001-35-2)
CI direct blue 15 (2429-74-5)	Melamine (108-78-1)	1,1,1-Trichloroethane (71-55-6)
Cinnamyl anthranilate (87-29-6)	Methoxychlor (72-43-5)	1,1,2-Trichloroethane (79-00-5)
Coumarin (91-64-5)	Methyl bromide (74-83-9)	Trichloroethylene (79-01-6)
m-Cresidine (102-50-1)	4,4'-Methylene bis(N,N-dimethyl)benzenamine (101-61-1)	1,2,3-Trichloropropane (96-18-4)
p-Cresidine (120-71-8)	Methyl methacrylate (80-62-6)	Triethanolamine (102-71-6)
D&C red 9 (5160-02-1)	4-Methyl-1-nitroanthraquinone (129-15-7)	Trifluralin (1582-09-8)
Dapsone (80-08-0)	N-Methylolacrylamide (924-42-5)	2,4,5-Trimethylaniline (137-17-7)
Decabromodiphenyl oxide (1163-19-5)	Methyl parathion (298-00-0)	Tris(2-chloroethyl)phosphate (115-96-8)
4,4'-Diaminodiphenyl ether (101-80-4)	Monuron (150-68-5)	Tris(2,3-dibromopropyl)phosphate (126-72-7)
1,4-Diamino-2-nitrobenzene (5307-14-2)	Naphthalene (91-20-3)	Vat yellow 4 (128-66-5)
1,2-Dibromo-3-chloropropane (96-12-8)	1,5-Naphthalenediamine (2243-62-1)	4-Vinylcyclohexene (100-40-3)
2,3-Dibromo-1-propanol (96-13-9)	Nithiazide (139-94-6)	4-Vinylcyclohexene diepoxide (106-87-6)
o-Dichlorobenzene (95-50-1)	Nitrilotriacetic acid (139-13-9)	Vinylidene chloride (75-35-4)
p-Dichlorobenzene (106-46-7)	5-Nitro-o-anisidine (99-59-2)	Vinyl toluene (25013-15-4)
1,2-Dichloroethane (107-06-2)	2-Nitroanisole (91-23-6)	Xylenes (1330-20-7)
Dichloromethane (75-09-2)	Nitrofen (1836-75-5)	Zidovudine (30516-87-1)
1,2-Dichloropropane (78-87-5)	Nitrofural (Nitrofurazone) (59-87-0)	Ziram (137-30-4)

[a]Information extracted from the *IARC Monograph* series and presented as agent (CAS number).
[b]NTP studies on 2,4-dinitrotoluene used.
Abbreviations: CI, color index; IARC, International Agency for Research on Cancer; NTP, National Toxicology Program.

position on potential human cancer and noncancer health effects that may result from chronic (or lifetime) oral or inhalation exposure to specific chemical substances in the environment. Combined with specific exposure assessment information, the summary health information in IRIS may be used as a source for the evaluation of potential public health risks from environmental contaminants (40).

The files on individual chemicals contain descriptive and quantitative information. EPA has relied upon findings and conclusions on chemicals from NTP toxicology and carcinogenicity studies, in calculating reference values (RfD and RfC)[b] and in reaching conclusions regarding the carcinogenicity of individual chemicals. Table 2 lists chemicals included in IRIS for which NTP studies are principal or supporting in the evaluation. There are 68 chemicals listed from 78 NTP technical reports. The majority of these chemicals (66) are detailed in the NTP Technical Report series on chronic exposure carcinogenicity studies.

NTP has conducted over 300 chronic exposure carcinogenicity studies since its inception in 1978; therefore, one might wonder why the other studies have had no impact on regulatory policy at EPA. Of those studies, 24% were negative, 19% were equivocal or inadequate, and 16% had findings in only one sex/species group (generally not sufficient weight-of-evidence for regulation), leaving 41% to be used by regulatory agencies. In many cases, the negative studies conducted by the NTP have led to no regulatory standard and this information would not appear in the IRIS database. Of the 41% that are positive, some were studies of food additives and/or industrial compounds not regulated by the EPA. Thus, approximately half of the positive studies conducted by the NTP have served as the primary experiments for regulations in the IRIS database and this is greater than 50% of the relevant studies.

IRIS database has 543 separate chemicals listed. Most of these compounds are proprietary substances for which U.S. law allows the EPA to require that testing be conducted by the manufacturer to evaluate safety. The NTP has provided primary data for about 12% of the compounds in the IRIS database and well over 50% of the data for nonproprietary substances.

Food and Drug Administration

Food and Drug Administration (FDA) relies upon data generated by the NTP in making decisions to safeguard public health, and the NTP facilitates the FDA's mission by conducting studies on agents of interest to the agency. Some of the current initiatives include NTP studies on ingredients in

---

[b] RfD is the reference dose for chronic noncarcinogenic health effects from oral exposure, and RfC is the reference concentration for chronic noncarcinogenic health effects from inhalation exposure (39).

Table 2 Chemicals for Which NTP Studies Were Used in Evaluations by the EPA[a]

Acetonitrile (75-05-8)	1,2-Dichlorobenzene (95-50-1)	Methyl methacrylate (80-62-6)
Acrolein (107-02-8)	1,4-Dichlorobenzene (106-46-7)	Methyl t-butyl ether[b] (1634-04-4)
Asbestos (1332-21-4)	1,1-Dichloroethylene[c] (75-35-4)	Mirex (2385-85-5)
Barium and compounds (7440-39-3)	Dichloromethane (75-09-2)	Monochloramine[d] (10599-90-3)
Bisphenol A (80-05-7)	1,2-Dichloropropane (78-87-5)	Naphthalene (91-20-3)
Boron (boron and borates only) (7440-42-8)[e]	1,3-Dichloropropene (542-75-6)	Pentachlorobenzene (608-93-5)
Bromodichloromethane (75-27-4)	Dichlorvos (62-73-7)	Pentachloronitrobenzene (82-68-8)
Bromoform (75-25-2)	Diethyl phthalate (84-66-2)	Pentachlorophenol (87-86-5)
Bromomethane (74-83-9)	1,2-Epoxybutane (106-88-7)	Phenol (108-95-2)
1,3-Butadiene (106-9-0)	Ethyl chloride (75-00-3)	Polychlorinated biphenyls (1336-36-3)[f]
Butyl benzyl phthalate (85-68-7)	Ethylbenzene (100-41-4)	Propylene oxide (75-56-9)
Caprolactam (105-60-2)	Ethylene glycol monobutyl ether[g] (111-76-2)	Pyridine (110-86-1)
Chloral hydrate (302-17-0)	Ethylene thiourea (96-45-7)	Quinoline[h] (91-22-5)
Chlorine (7782-50-5)	Fluometuron (2164-17-2)	Rotenone (83-79-4)
2-Chloroacetophenone (532-27-4)	Furan (110-00-9)	Selenium sulfide (7446-34-6)

Chlorobenzene (108-90-7)	Furfural (98-01-1)	Sodium azide (26628-22-8)
1-Chlorobutane (109-69-3)	Hexachlorocyclopentadiene (77-47-4)	Styrene (100-42-5)
Decabromodiphenyl ether (1163-19-5)	Hexachlorodibenzo-p-dioxin, mixture (19408-74-3)	1,1,1,2-Tetrachloroethane (630-20-6)
Di(2-ethylhexyl)adipate (103-23-1)	Hexachloroethane (67-72-1)	Tetrachloroethylene (127-18-4)
Di(2-ethylhexyl)phthalate (117-81-7)	Isophorone (78-59-1)	Toluene (108-88-3)
1,2-Dibromo-3-chloropropane (96-12-8)	D-Limonene (5989-27-5)	1,2,3-Trichloropropane (96-18-4)
Dibromochloromethane (124-48-1)	Mercuric chloride (7487-94-7)	Xylenes (1330-20-7)
1,2-Dibromoethane (106-93-4)	Methyl methacrylate (80-62-6)	

[a]Information extracted from the EPA IRIS database (http://www.epa.gov/iris) and presented as chemical (CAS number).
[b]NTP studies on dimethyl methylphosphonate used.
[c]NTP studies on vinylidene chloride used.
[d]NTP studies on chloraminated water used.
[e]NTP studies on boric acid used.
[f]NTP studies on aroclor, 2,3,7,8-tetrachlorodibenzo-p-dioxin, and polybrominated biphenyl mixture used.
[g]NTP studies on 2-butoxyethanol used.
[h]NTP studies on 8-hydroxyquinoline used.
Abbreviations: EPA, Environmental Protection Agency; IRIS, Integrated Risk Information System; NTP, National Toxicological Program.

sunscreens and cosmetics (see section "Phototoxicology"), herbal medicines (see section "Medicinal Herbs"), and radiofrequency radiation emissions from cellular telephones (see below).

An example of where the FDA has used information from the NTP is in the evaluation of proposed food additives. Under the general safety standard of the Federal Food, Drug, and Cosmetic Act [21 U.S.C. 348(c)(3)(A)], the FDA must evaluate the data provided on food additives to determine whether they are safe for their intended use. The FDA reviews the safety of the additive itself and of any chemical impurities that may be present in the additive as a result of owing to its manufacturing process. The FDA used NTP chronic toxicology and carcinogenicity studies in its risk evaluations of the following carcinogenic impurities in food additives:

- *p*-chloroaniline hydrochloride, an impurity in the food additive 2,9-dichloro-5,12-dihydroquinone[2,3-*b*]acridine-7,14-dione (C.I. Pigment Red 202) intended for use as a colorant in polymers in contact with food (41);
- tetrachloroethylene, an impurity in the food additive 4,5-dichloro-3H-1,2-dithiol-3-one intended for use as a slimicide in the manufacture of food-contact paper and paperboard (42);
- benzene, an impurity in ethylene–norborene copolymers intended as articles or components in contact with dry food (43).

NTP has studied a number of pharmaceuticals and food agents, including some nominated by the FDA—the antihistamine chlorpheniramine maleate, the sedative chloral hydrate, the tranquilizer oxazepam, the laxative ingredient senna, the orphan drug Elmiron® (Ortho-McNeil Pharmaceutical, Inc., New Jersey, U.S.), and the flavoring agent cinnamaldehyde. NTP studies contributed to FDA's decision in 1999 to reclassify phenolphthalein, an ingredient in over-the-counter laxative drugs, to "not generally recognized as safe and effective" [21 CFR 310.545(a)(12)(iv)(B)] (44).

Current NTP Initiatives of Priority to Regulatory Agencies

Radiofrequency Radiation Emissions from Cellular Phones

More than 100 million people in the United States currently use wireless communication devices, with thousands of new users added daily. Personal (cellular) telecommunications is a rapidly evolving technology that uses microwave radiation to communicate between a fixed base station and a mobile user. Most systems employ a hand-held cellular telephone, with the radiation antenna held close to the user's head. The Federal Communication Commission requires cellular phones and other wireless communication devices to meet its guidelines for exposure to microwave radiation. These guidelines are based on the need to protect the user from immediate injury from the heat produced by microwave radiation.

FDA nominated radiofrequency radiation emissions from cellular telephones to the NTP for study because data available at present are insufficient to determine whether current guidelines are adequate for protecting against potential adverse effects of long-term exposure. Studies in laboratory animals are crucial for understanding whether exposure to microwave radiation may pose a danger to human health. Other research groups in Europe and Australia are performing several long-term animal studies addressing this issue; however, the NTP plans to conduct laboratory research to help clarify any potential health hazard. NTP will study the toxic and carcinogenic effects of chronic exposure to cell phone microwave radiation emissions in laboratory animals. NTP has worked with technical experts from the National Institute of Standards and Technology (NIST) to test the suitability of various microwave radiation exposure systems for these studies. Based upon NIST studies, a reverberation chamber exposure system will be used and the NTP laboratory studies were anticipated to get underway in 2005 (9).

Safe Drinking Water

More than 200 million people in the United States are estimated to use municipally treated drinking water; hence the availability of safe drinking water is of enormous importance to public health. Although chlorination is one of the major public health advances of the 20th century, by-products of chlorination or other disinfection processes [disinfection by-products (DBPs)] may cause health problems such as cancer. Moreover, some agents found naturally in water or that contaminate public water systems may pose a threat to public health.

EPA is responsible for setting water standards for DBPs. To provide scientific data for setting sound standards for water quality, the NTP is collaborating with the EPA on a research program to assess potential risks from human exposure to DBPs. This program includes a systematic, mechanism-based evaluation of DBPs focusing on reproductive toxicity, immunotoxicity, neurotoxicity, and carcinogenicity. The program is selecting DBPs for study, based on their presence in drinking water, occurrence with different disinfection processes, chemical structures, and class (trihalomethanes, haloacetic acids, and haloacetonitriles).

In response to concerns by a number of California legislators, the California EPA, and the California Health and Human Services Agency, the NTP is studying the potential of hexavalent chromium, added to drinking water, to cause cancer. Chromium is a naturally occurring element present in various valence states. Trivalent chromium is an essential nutrient, and in nature, chromium occurs most commonly in this state. Hexavalent chromium compounds are the next most stable form; however, they rarely occur naturally and are typically associated with industrial sources.

Hexavalent chromium is an established human carcinogen in certain occupational settings, presumably as a result of inhalation exposure.

However, the long-term consequence of exposure to hexavalent chromium compounds in the water supply is not known. Data currently available on the chronic toxicity and carcinogenicity of hexavalent chromium given orally are not sufficient to establish or characterize any hazard. NTP studies include both short- and long-term administration of hexavalent chromium as sodium dichromate dihydrate, in drinking water, to laboratory animals. California EPA will use information from these studies to help set permissible limits for hexavalent chromium in drinking water (9).

Medicinal Herbs

Medicinal herbs are among our oldest medicines, and their increasing use in recent years is evidence of public interest in alternatives to conventional medicine. About one-third of the U.S. population is believed to use some form of alternative medicine, including herbal remedies. The use of herbal medicines and other dietary supplements has increased substantially since passage of the 1994 Dietary Supplement Health and Education Act (Public Law 103–417). Although about 1500 botanicals are sold as dietary supplements or ethnic traditional medicines, according to provisions in the Act, herbal formulations are not subjected to FDA premarket approval to ensure their safety or efficacy. Unfortunately, medicinal herbs are poorly defined mixtures and the ingredients, both active and otherwise, may vary greatly from one preparation to another. There is a need for efforts to assure the quality, efficacy, and safety of medicinal herbs.

NTP is planning to conduct research on several medicinal herbs and compounds found in herbs to examine carcinogenicity, reproductive toxicity, neurotoxicity, immunotoxicity, or toxic effects associated with exposures to high acute doses and chronic low doses. These herbal medicines include aloe vera gel, black cohosh, black walnut extract, comfrey, *Echinacea purpurea* extract, ephedra, *Ginko biloba* extract, goldenseal powder, grape seed extract, pine bark extract, kava kava, milk thistle extract, pulegone, senna, and thujone (9). Pyrrolizidine alkaloids are known hepatotoxins in animals. NTP's results on the pyrrolizidine alkaloid content of comfrey led the FDA to issue an advisory on July 6, 2001, to manufacturers to remove dietary supplements containing comfrey from the market and alert customers to stop using the product (45).

Nanoscale Materials

Nanotechnology, in recent years, has become an increasing focus of U.S. and global research and development efforts. As with many technological advances, new materials are created, and as a result, the potential exists for new and unanticipated human exposures for which the human health impact is unknown. NTP is developing a broad-based research program to address potential human health hazards associated with the manufacture and use of nanoscale materials. This research program will include studies of nanoscale materials that apply existing toxicology testing methods and also explore the development of appropriate novel toxicological methods to adequately assess potential human health effects.

Nanoscale materials are a broadly defined set of substances where at least one critical dimension is less than 100 nm. Ultrafine particulate matter is a well-known example of ambient nanoscale particles; however, the NTP's research program will initially focus on manufactured nanoscale materials of current or projected commercial importance. Nanoscale materials can, in theory, be engineered from nearly any chemical substance; semiconductor nanocrystals, organic dendrimers, and carbon fullerenes and carbon nanotubes are a few of the many examples. They are already appearing in commerce as industrial and consumer products and as novel drug delivery formulations.

The intent of the NTP's research program is to evaluate the toxicological properties of major nanoscale materials classes which represent a cross-section of composition, size, surface coatings, and physicochemical properties, and to use these as model systems to investigate fundamental questions concerning if and how nanoscale materials can interact with biological systems. Some of these fundamental questions are as follows: What are the appropriate methods for detection and quantification of nanoscale particles in tissues? How are nanoscale materials absorbed, distributed in the body, and taken up by cells? Are there novel toxicological interactions? As part of this research program, studies to evaluate the biological disposition of nanoscale crystalline fluorescent semiconductors ("quantum dots"), long-term toxicology studies of carbon-based nanoscale materials (e.g., single- or multiwalled nanotubes, fullerenes), and phototoxicology studies of representative nanoscale metal oxide particles used in industrial settings and consumer products (e.g., titanium dioxide) are being considered (9).

Phototoxicology

The U.S. public is increasingly exposed to ultraviolet (UV) radiation from sunlight owing to more leisure time spent in outdoor activities, and also from other sources (e.g., tanning booths). NTP is coordinating an effort between the NIEHS and NCTR to study the phototoxicology and photocarcinogenicity of substances nominated to the NTP, including those of high priority to the FDA. In general, these studies investigate the effects on gene expression, toxicity, and carcinogenicity of sunlight combined with either topically or systemically applied substances in the SKH-1 hairless mouse. Much of this research is being carried out at the NTP Center for Phototoxicology at the NCTR in Jefferson, Arkansas.

Phototoxicology studies for several topically applied compounds are in progress. Many cosmetics include alpha-hydroxy and beta-hydroxy acids as chemical exfoliating agents to correct or improve the appearance of "sun-aged" skin. The relation of skin cancer to their continuous use combined with exposure to sunlight is not known and is currently under study. Studies are also investigating the possible acute toxicity and carcinogenicity of topically applied plant fractions of the aloe vera plant or topically applied retinyl palmitate in combination with simulated sunlight. Many products, including cosmetics and dietary supplements, contain portions of the aloe vera plant, and retinyl palmitate is included in some cosmetics as an "anti-wrinkle" compound. In addition, other cosmetic ingredients including nanoscale particles used in sunscreens (zinc oxide and titanium dioxide) will be studied in the future (9).

Endrocrine-Disrupting Agents

Endocrine disruptors are naturally occurring or synthetic substances that may mimic or interfere with the natural hormones in the body. Endocrine disruptors may turn on, turn off, or change signals that hormones carry and thus affect the normal functions of tissues and organs. NTP is involved in several efforts to strengthen the scientific knowledge within this field.

Endocrine-disrupting chemicals are of interest to the FDA, and the NTP is coodinating an effort between the NIEHS and NCTR to conduct toxicology studies on chemicals that include the plant-based estrogen (phytoestrogen) genistein, the pesticide vinclozolin, the drug ethinyl estradiol, and the industrial chemical nonylphenol. These studies assess effects on reproduction, development of hormone-sensitive organs, cancer in rodents over several generations, and behavioral and immunological effects (9).

Occupational Exposures

NTP is coordinating an effort between the NIEHS and NIOSH to better understand worker exposures, educate workers, and identify occupational health research gaps. A current effort is addressing worker exposure to welding fumes and 1-bromopropane. Studies of diseases in workers suggest that occupational exposure to welding fumes may cause adverse health effects. More information is needed to evaluate the relationship between timing and amount of exposure and the adverse effects, and to understand the specific causes of these effects. NIOSH has constructed a computer-controlled, automated robotic welding fume inhalation system and will characterize the physical and chemical compositions of the generated fumes and gases. In addition, studies will be performed to evaluate the exposure conditions, generator parameters, and welding processes and materials that cause acute responses in laboratory animals by assessing lung injury, inflammation, and changes in the immune system.

An industry consortium petitioned the EPA to list 1-bromopropane as an alternative for ozone-depleting solvents. This could vastly increase the exposure of workers and the public to this compound. To obtain information on exposures to this chemical, NIOSH conducted an industry-wide study targeting industries that use adhesives, the metal degreasing and electronics industry, and chemical, aerosol, and adhesive manufacturers. The NTP is also conducting laboratory studies to evaluate the potential toxicity and carcinogenicity of 1-bromopropane (9).

Use of Findings from NTP Toxicology and Carcinogenicity Studies in Proposition 65 Listings

The Safe Drinking Water and Toxic Enforcement Act of 1986 ("The Act"), also known as Proposition 65, is a broad statute that requires the Governor of California to revise and publish, at least annually, a listing of chemicals known to the State of California to cause cancer or reproductive toxicity (California Health and Safety Code, Section 25249.5 et seq; Title 22, California Code of Regulations, Section 12000 et seq.) (46,47). The Act prohibits discharges of specific agents into sources of drinking water and requires that warnings be given to individuals exposed to these agents in the workplace, environment, or through consumer products unless the discharges or exposures pose no significant health risks (46). The Office of Environmental Health Hazard Assessment (OEHHA) of the California EPA is the lead agency for implementation of Proposition 65 [Title 22, California Code of Regulations, Section 12102(o)].

There are three principal ways for a chemical to be added to the Proposition 65 list. The three mechanisms are (*i*) if the "state's qualified experts" find that the chemical has been clearly shown "through scientifically valid testing... to cause cancer or reproductive toxicity," (*ii*) if an "authoritative body" formally identifies it as causing cancer or reproductive toxicity, or (*iii*) if an agency of the state or federal government requires that it be "labeled or identified as causing cancer or reproductive toxicity" [California Health and Safety Code, Section 25249.8(b)]. Members of the Carcinogen Identification Committee (CIC) and the Developmental and Reproductive Toxicant (DART) Identification Committee are considered the "state's qualified experts" for rendering opinion about whether a chemical has been clearly shown to cause cancer or reproductive toxicity, respectively, and for identifying authoritative bodies under Proposition 65 (Title 22, California Code of Regulations, Section 12305).

Under the first mechanism, the CIC or DART Identification Committee reviews all relevant scientific literature compiled by OEHHA, considers comments from the public, and renders a decision about the placement of a chemical on the list. NTP technical reports have contributed to the body of literature on chemicals reviewed by the CIC for possible listing as causing cancer under Proposition 65 (46).

Under the second mechanism, the CIC or DART Identification Committee designates specific groups as "authoritative bodies" for identification of chemicals as causing cancer or reproductive toxicity, respectively. The regulatory guidance for evaluating the documentation and scientific findings for listing chemicals through the "authoritative bodies" mechanism is provided in Title 22, California Code of Regulations, Section 12306. The CIC has identified the NTP as an authoritative body for identifying chemicals known to cause cancer under Proposition 65 in addition to other groups, including the EPA, IARC, the FDA, and NIOSH [Title 22, California Code of Regulations, Section 12306(m)].

OEHHA examines documents released by authoritative bodies to determine whether they satisfy the requirements for possible listing of chemicals under Proposition 65 via the authoritative bodies mechanism. OEHHA has formally reviewed documents produced by the NTP and determined that the NTP Technical Report series and the RoC (described below) satisfy the regulatory criteria for "formal identification" of a chemical as causing cancer and may serve as the basis for an "authoritative bodies" listing under Proposition 65 [Title 22, California Code of Regulations, Section 12306(d)] (48).

OEHHA reviews individual NTP technical reports with findings of "clear evidence" of carcinogenic activity in at least one experiment and determines whether listing via the authoritative bodies mechanism is required (48). Specifically, OEHHA determines whether the NTP concludes that the chemical causes cancer and whether the evidence provided meets the technical criteria for "causing cancer" [Title 22, California Code of Regulations, Section 12306(e)] for listing under Proposition 65. The evidence from the NTP's laboratory studies in animals is considered "sufficient" for listing if the studies "indicate that there is an increased incidence of malignant tumors or combined malignant and benign tumors in multiple species or strains, in multiple experiments (e.g., with different routes of administration or using different dose levels), or, to an unusual degree, in a single experiment with regard to high incidence, site or type of tumor, or age at onset" [Title 22, California Code of Regulations, Section 12306(e) (2)].

Currently, 478 chemicals are listed as known to the State of California to cause cancer and 165 of those listings occurred by the authoritative bodies mechanism (C. Oshita, personal communication, 2004). The list, which is accessible from the OEHHA web site (49), includes naturally occurring or synthetic chemicals, such as additives or ingredients in pesticides, common household products, drugs, dyes or solvents, and chemicals that are used in manufacturing or construction or are by-products of chemical processes (46). Findings from NTP technical reports have served as the sole or partial basis for the listing of approximately one-fourth (48) of the 165 chemicals via the authoritative bodies mechanism as causing cancer (Table 3). In addition, NTP studies were reviewed in IARC evaluations for 6 of the 67 chemicals listed under Proposition 65 recognizing IARC as an "authoritative body."

Table 3 Chemicals Listed Under Proposition 65 as Known to Cause Cancer via the "Authoritative Bodies" Mechanism Based on NTP Technical Reports[a]

1-Amino-2,4-dibromoanthraquinone (81-49-2)	3,3'-Dimethylbenzidine dihydrochloride (612-82-8)	o-Nitrotoluene (88-72-2)
Benzofuran (271-89-6)	Fumonisin B1[b] (116355-83-0)	Ochratoxin A (303-47-9)
2,2-bis(Bromomethyl)-1,3-propanediol (3296-90-0)	Furan (110-00-9)	Oxazepam (604-75-1)
Bromoethane (74-96-4)	Glycidol (556-52-5)	Pentachlorophenol[c] (87-86-5)
C.I. Acid Red 114 (6459-94-5)	Hexachloroethane[d] (67-72-1)	Phenolphthalein (77-09-8)
C.I. Direct Blue 218 (28407-37-6)	Indium phosphide (22398-80-7)	Primidone (125-33-7)
C.I. Solvent Yellow 14 (842-07-9)	Isobutyl nitrite (542-56-3)	Pyridine (110-86-1)
p-Chloroaniline hydrochloride (20265-96-7)	Methyl carbamate (598-55-0)	Salicylazosulfapyridine (599-79-1)
Chloroethane (75-00-3)	Methyleugenol (93-15-2)	Tetrafluoroethylene (116-14-3)
Chloroprene[e] (126-99-8)	N-Methylolacrylamide (924-42-5)	Tetranitromethane (509-14-8)
Cobalt sulfate heptahydrate (10026-24-1)	Nalidixic acid (389-08-2)	1,2,3-Trichloropropane (96-18-4)
Cytembena (21739-91-3)	Naphthalene (91-20-3)	Tris(2-chloroethyl) phosphate (115-96-8)
D&C Red No. 9[f] (5160-02-1)	o-Nitroanisole (91-23-6)	4-Vinyl-1-cyclohexene diepoxide (106-87-6)
2,3-Dibromo-1-propanol (96-13-9)	Nitromethane (75-52-5)	2,6-Xylidine (87-62-7)
3,3'-Dimethoxybenzidine dihydrochloride (20325-40-0)		

[a]Listing presented as CAS number.
[b]Listing based upon NTP and IARC as authoritative bodies.
[c]Listing based upon NTP and EPA as authoritative bodies.
[d]Listing based upon NTP and NIOSH as authoritative bodies.
[e]Listing based upon NTP and IARC as authoritative bodies.
[f]Listing based upon NTP and FDA as authoritative bodies.

Abbreviations: CI, color index; EPA, Environmental Protection Agency; FDA, Food and Drug Administration; IARC, International Agency for Research on Cancer; NIOSH, National Institute for Occupational Safety and Health; NTP, National Toxicological Program.
Source: C. Oshita, personal communication, 2004.

REPORT ON CARCINOGENS

In response to public concern about the relationship between the environment and cancer, the Congress mandated the Secretary of Health and Human Services (HHS) to publish a list of agents that are known or reasonably anticipated to be human carcinogens, and to which a significant number of people in the United States are exposed [42 U.S.C Section 241(b)(4) as amended]. The report changed from an annual to biennial publication in 1993 (50,51).

The RoC is an informational, scientific, and public health document that discusses and identifies substances (including agents, mixtures, chemicals, or exposure circumstances) that may pose a carcinogenic hazard to human health. The responsibility for preparation of the RoC is assigned to the NTP. The review of nominations for listing in or delisting (removal) from the RoC follows a formal process that includes many phases of scientific peer review and many opportunities for public comment. Scientific review groups evaluate each nomination according to specific criteria. A recommendation for listing in or delisting from the report is based on the strength of the evidence for carcinogenicity. The NTP Director evaluates all review group recommendations, public comments, and other information and formulates a recommendation to the Secretary, HHS, on the disposition of each nomination. Following the Secretary's approval, the RoC is forwarded to the Congress and made available to the public (52).

The RoC is required to identify each listed substance for which no standard on exposure or release into the environment has been established by a federal agency. The report also contains, for each listing, a summary of federal regulations; however, in some cases, these standards and regulations have been enacted for reasons other than the carcinogenicity of the substance, for instance, to prevent other illness or to improve the quality of food or the environment (52).

Use of Listings in the RoC in Public Health

As a vehicle for communicating information about carcinogenic hazards, listings in the RoC provide a basis to help state and federal regulatory agencies identify priorities (52,53), and the report serves as an important mechanism for alerting the public about potential cancer threats (6). This section highlights some examples where the RoC initiates decisions in regulations or state legislation.

Occupational Safety and Health Administration

The OSHA Hazard Communication regulation (Title 29, Code of Federal Regulations, Section 1910.1200) is intended "to ensure that the hazards of all chemicals produced or imported are evaluated and that information concerning their hazards is transmitted to employees and employers."

This transmittal of information can occur through comprehensive hazard communication programs that include container labeling and other forms of warning, materials safety data sheets, and employee training about the hazardous nature of the chemicals and appropriate protective measures. The regulation [Subsection 1200(d) (4) and Appendix A] recognizes the IARC and the RoC as sources for establishing that a chemical is a carcinogen or a potential carcinogen for hazard communication purposes. In this way, the NTP impacts worker safety and health directly through the listing of chemical hazards in the RoC and indirectly through the conduct of toxicology and carcinogenicity studies whose findings are frequently used in IARC evaluations.

OSHA also uses NTP data for setting occupational standards. For example, OSHA considered NTP studies as part of the data used to support the amending of the standard that regulates worker exposure to 1,3-butadiene (48) and methylene chloride (55).

California Labor Law

The California Labor Code (Section 6380) provides for establishment of a list of hazardous substances that is available to manufacturers, employers, and the public. Section 6382(b) (1) provides for listing chemicals recognized as human or animal carcinogens by IARC and Section 6382(d) refers to chemicals recognized as carcinogens within the scope of the federal Hazard Communications Standard (Title 29, Code of Federal Regulations, Section 1910.1200). As noted above, the Hazard Communications Standard recognizes the RoC as a source for identification of carcinogens and potential carcinogens; therefore, it contributes indirectly to the identification of chemicals as occupational hazards in California. In addition, NTP studies contribute to the body of knowledge used by IARC to evaluate whether substances are carcinogenic to humans and, therefore, indirectly impact the identification of hazardous substances.

Proposition 65

The RoC influences the listing of chemicals known to cause cancer under Proposition 65 in at least two ways. First, substances listed in the RoC are subject to listing under Proposition 65 by cross-reference to Section 6382(d) of the California Labor Code in Section 25249.8 of the Safe Drinking Water and Toxic Substances Act (53). Second, the RoC can trigger possible listing under Proposition 65 directly through recognition of the NTP as an "authoritative body" for identification of chemicals causing cancer [Title 22, California Code of Regulations, Section 12306(m)]. Interestingly, the RoC has affected both the listing and delisting of chemicals under Proposition 65.

The RoC has contributed to the listing of approximately one-third of the 478 chemicals listed as carcinogens under Proposition 65 as of

November 2003. By reference to the California Labor Code, Section 6382(d), 25 of 26 chemicals listed initially under Proposition 65 on February 27, 1987, were added based solely, or in part, on their designation in the RoC as known to be human carcinogens (L. Zeise, personal communication, 2004). Subsequent to a judicial decision interpreting Labor Code Sections 6382(b)(1) and (d) to include chemicals recognized as carcinogens in animals on the Proposition 65 list, 15 chemicals were added on October 1, 1989, based solely, or in part, on their designation in the report as reasonably anticipated to be human carcinogens (L. Zeise, personal communication, 2004) (56). In addition, in initial efforts to list chemicals under Proposition 65 and in the absence of identified authoritative bodies, the Scientific Advisory Panel (now replaced by the CIC) served as the "state's qualified experts" and evaluated chemicals identified as carcinogens by IARC and the RoC. Listing in the RoC contributed to the panel's decision to list approximately 130 chemicals under Proposition 65 (L. Zeise, personal communication, 2004).

Changes in the listing of chemicals by IARC and the RoC can potentially impact their listing under Proposition 65. Nickel and certain nickel compounds were listed by reference to the California Labor Code Section 6382(b) (1) and (d) in October 1, 1989. Subsequently, IARC and NTP updated their listing designations from "nickel and certain nickel compounds" to "metallic nickel" and "nickel compounds." OEHHA clarified the listing of nickel and certain nickel compounds on June 6, 2003 (57) and amended the Proposition 65 list on May 7, 2004, to include nickel compounds (58).

Of the 478 chemicals recognized under Proposition 65 as known to the State of California to cause cancer, two were listed via the authoritative bodies mechanism (Title 22, California Code of Regulations, Section 12306) based on being listed in the RoC (C. Oshita, personal communication, 2004).

- 3,3'-Dichlorobenzidine dihydrochloride was listed under Proposition 65 on May 15, 1998. It was first listed as *reasonably anticipated to be a human carcinogen* in the second edition of the RoC (59).
- Strong inorganic mists containing sulfuric acid were listed on March 14, 2003. Strong inorganic acid mist containing sulfuric acid was first listed as a *known human carcinogen* in the ninth edition of the RoC (60).

Although chemicals are generally added to the Proposition 65 list, at least one chemical has been removed. OEHHA removed saccharin (CAS No. 81-07-2) from the list of chemicals known to the State of California to cause cancer for purposes of Proposition 65 on April 6, 2001. It was originally added to the Proposition 65 list on October 1, 1989, based upon its identification as a carcinogen by IARC and NTP through enforcement of the California Labor Code 6382(b) (1) and (d), which provides for listing under Proposition 65 pursuant to the California Health and Safety Code Section 25249.8(a) (see above). In 1999, the IARC reclassified saccharin and its salts as Group 3, *not classifiable to their carcinogenicity to humans* (61). The NTP removed saccharin from the RoC, ninth edition, based upon formal review and approval for delisting by the Secretary, HHS (60). Because saccharin was added to the Proposition 65 list based upon the Labor Code Section that references chemicals identified as known or potential carcinogens by IARC or NTP, the removal of this designation by these groups led to saccharin being removed from the Proposition 65 list (62).

CENTER FOR THE EVALUATION OF RISKS TO HUMAN REPRODUCTION

Established in 1998, the NTP's Center for the Evaluation of Risks to Human Reproduction (CERHR) serves as a resource to the public, health and regulatory agencies, and the medical and scientific communities. The CERHR is unique in being the only program of this type. It provides uniform and scientifically based assessments of the potential for adverse effects on reproduction and development caused by agents to which humans are exposed in our environment. The CERHR carries out its assessments, which include scientific peer review by independent scientific experts, in public forums with opportunity for public comment. The panel evaluates a chemical's potential reproductive and/or development toxicity according to established guidelines accessible on the CERHR web site (63). By the end of 2004, CERHR expert panels had addressed the potential reproductive and/or developmental toxicity of seven phthalates,[c] methanol, 1-bromopropane, 2-bromopropane, ethylene glycol, propylene glycol, acrylamide, and fluoxetine (9). Once an expert panel review is completed, the CERHR produces an NTP-CERHR monograph that includes the expert panel report, all public comments received on the expert panel report, and the NTP brief. The NTP brief provides the NTP's interpretation of the potential for the chemical to cause adverse reproductive and/or developmental effects to humans exposed to it. As and when completed, expert panel reports and NTP-CERHR monographs are posted on the CERHR web site (7).

In the short time since its establishment, CERHR's evaluations of the potential reproductive and development toxicity of chemicals are playing a major role in public health decision making. At present, most effects are indirect, in cases where CERHR expert panel reports serve as background information to an agency's or panel's review. Some examples where

---

[c] Butyl benzyl phthalate, di(2-ethylhexyl) phthalate, di-isodecyl phthalate, di-isononyl phthalate, di-*n*-butyl phthalate, di-*n*-hexyl phthalate, and di-*n*-octyl phthalate.

CERHR expert panel reports influenced evaluations by health and regulatory agencies include the following:

- Health Canada's Therapeutics Products Directorate Expert Advisory Panel on di(2-ethylhexyl)phthalate (DEHP) in Medical Devices included the *NTP-CERHR Expert Panel Report on DEHP* (64) in its assessment of the potential reproductive risk associated with exposure to this chemical. The Therapeutic Products Directorate is the Canadian federal authority that regulates pharmaceuticals and medical devices for human use. Conclusions reached by the expert advisory panel regarding the reproductive and developmental toxicity of DEHP were similar to those of the CERHR expert panel (65).
- As part of an investigation of diisonyl phthalate (DINP), a replacement for DEHP in teethers, rattles, and pacifiers, the CPSC convened the Chronic Hazard Advisory Panel (CHAP) "to determine whether DINP is a carcinogen, mutagen, or teratogen or poses some other chronic hazard, and if feasible, estimate the probable harm to human health that will result from exposure to DINP" (66). The *NTP-CERHR Expert Panel Report on DINP* (67) contributed signficantly to the CHAP's discussions and relevant sections were reproduced with minor changes in the CHAP report. The CHAP's conclusions for effects of DINP on reproduction and development were similar to those reached by the CERHR expert panel (66).
- The EPA referenced the CERHR evaluation of 1-bromopropane (68) (also known as *n*-propyl bromide) in its proposed rule for the use of 1-bromopropane as an acceptable chemical alternative to ozone-depleting substances under the EPA's Significant New Alternatives Policy (SNAP) Program published in the *Federal Register* on June 3, 2003 (69).
- In considering an acceptable level of isopropyl bromide (also known as 2-bromopropane, CAS No. 75-26-3) contamination in *n*-propyl bromide formulations, the EPA considered the *NTP-CERHR Expert Panel Report on 2-Bromopropane* (70) in its evaluation of the reproductive toxicity of isopropyl bromide. Based upon the available information, the EPA concluded that it is appropriate to limit the amount of isopropyl bromide exposure from *n*-propyl bromide use (69).

Proposition 65

As noted previously, Proposition 65 requires the listing of chemicals known to the State of California to cause reproductive toxicity. The DART Identification Committee assists the lead agency, OEHHA, in the identification of "authoritative bodies" for formally identifying chemicals as causing reproductive toxicity [Title 22, California Code of Regulations, Section 12305(b)(2)]. The regulatory guidance for evaluating the documentation and scientific findings for listing chemicals through the authoritative bodies mechanism is provided in Section 12306. One mechanism for formally identifying a chemical as causing reproductive toxicity is the publishing of a report concluding that the chemical causes reproductive toxicity, by the authoritative body [22 CCR Section 12306(d)(1)]. As discussed above, the NTP is designated an authoritative body for identifying chemicals as causing cancer.

NTP was considered an authoritative body for identification of chemicals as reproductive toxicants until July 27, 1998, when the DART Identification Committee de-designated it (71). Prior to this only one chemical, nitrofurantoin, had been added to the Proposition 65 list based on the NTP (C. Oshita, personal communication, 2004). The Committee noted that it would reconsider this decision after the NTP established the CERHR (71). On December 4, 2002, the DART Identification Committee redesignated the NTP as an authoritative body "solely as to final documents released by the NTP's Center for the Evaluation of Risks to Human Reproduction" (72). Section 12306(l) of Title 22, California Code of Regulations, was amended to reflect this designation in February 19, 2003 (C. Oshita, personal communication, 2004) (73). This redesignation is already having an impact as OEHHA announced in 2004 consideration of possible listing of 1-bromopropane, butyl benzyl phthalate, di-*n*-butyl phthalate, di-*n*-hexyl phthalate, and di-isodecyl phthalate as causing reproductive toxicity via the authoritative bodies mechanism (74).

SUMMARY

Since its inception in 1978, the NTP has served as a focal point for coordinating toxicological testing programs within the DHHS and strengthening the science base used in public health decision making. The program works to make its research and testing program responsive to the needs of health and regulatory agencies, and to provide them with the critical data needed for making informed decisions that protect public health and the environment. Currently, the NTP has on-going studies in several areas that are of interest to its federal agency partners and have received inadequate attention in the past, such as photoactive chemicals, contaminants of finished drinking water, and endocrine-disrupting agents, and is addressing the potential safety issues associated with herbal medicines, radiofrequency radiation emissions from cellular telephones, and occupational exposures. In general, these initiatives are broad based and focused on investigating health-related effects associated with exposure to the chemical or physical agent.

NTP fosters public debate on issues of public health concern and encourages involvement by all interested parties in those discussions. The program also works to maintain transparency and openness in its processes to ensure that all issues are being addressed and the best scientific information is considered. Estimating the actual impact of the NTP in public health decision-making is beyond the scope of this chapter; however, we have attempted to highlight some examples where programmatic information is being used nationally and internationally to show the breadth of the NTP's role. Findings from NTP studies reported as NTP technical reports provide unique, peer-reviewed, and publicly available information used by various groups in assessing the carcinogenic hazard of chemical and physical agents. The RoC and the CERHR, which each evaluate the strength of the scientific knowledge in their assessment of an agent, attempt to place this information in a context useful for assessing potential hazard for humans. As shown in the examples provided, the RoC and NTP technical reports have direct and indirect impacts on regulations and state legislation. Evaluations conducted to date by the CERHR have been well received by other agencies and are beginning to have an influence on public health decisions; this impact is likely to increase as the CERHR evaluates more chemicals for their potential reproductive and/or developmental effects on human health.

In summary, the NTP was established to meet scientific, regulatory, and Congressional concerns about the health effects of chemical agents in our environment and has been highly successful in meeting the needs of its audience. The program will continue to strive to remain at the cutting edge of scientific research and the development and application of new technologies, so that the best science is available for reaching the best decisions that protect human health and the environment.

REFERENCES

1. Fisher BE. Spheres of influence: 20 years of toxicology. Environ Health Perspect 1998; 106(10):A484–A487.
2. HEW. Establishment of a National Toxicology Program. Federal Register 1978; 43(221):53,060–53,061.
3. National Toxicology Program Annual Plan for Fiscal Year 1983. Research Triangle Park: National Toxicology Program, Department of Health and Human Services, U.S. Public Health Service, 1983.
4. National Toxicology Program Fiscal Year 1982 Annual Plan. Research Triangle Park: National Toxicology Program, Department of Health and Human Services, U.S. Public Health Service, 1982.
5. NTP Current Directions and Evolving Strategies. Research Triangle Park: National Toxicology Program, Department of Health and Human Services, U.S. Public Health Service, 2001.
6. Lucier G, Barrett JC. Public health policy and the National Toxicology Program. Environ Health Perspect 1998; 106(10):A470–A471.
7. NTP Current Directions and Evolving Strategies. Research Triangle Park: National Toxicology Program, Department of Health and Human Services, U.S. Public Health Service, 2002.

8. Chhabra RS, Bucher JR, Wolfe M, Portier C. Toxicity characterization of environmental chemicals by the US National Toxicology Program: an overview. Int J Hyg Environ Health 2003; 206(4–5):437–445.
9. NTP Current Directions and Evolving Strategies. Research Triangle Park: National Toxicology Program, U.S. Department of Health and Human Services, Public Health Service, 2004.
10. Scientific Issues Relevant to Assessment of Health Effects from Exposure to Methylmercury. Raleigh, NC: National Institute of Environmental Health Sciences, November 18–20, 1998.
11. NIEHS news: herbal health. Environ Health Perspect 1999; 106(12):A590–A592.
12. Matthews HB, Lucier GW, Fisher KD. Medicinal herbs in the United States: research needs. Environ Health Perspect 1999; 107:773–778.
13. The Role of Human Exposure Assessment in the Prevention of Environmental Disease. Rockville, MD: National Institute of Environmental Health Sciences, 1998.
14. Final Report of the Endocrine Disruptors Low-Dose Peer Review. Research Triangle Park, NC: National Institute of Environmental Health Sciences, 2001.
15. Melnick R, Lucier G, Wolfe M, et al. Summary of the National Toxicology Program's report of the endocrine disruptors low-dose peer review. Environ Health Perspect 2002; 110(4):427–431.
16. Selgrade MJK, Kimber I, Goldman L, Germolec DR. Assessment of allergenic potential of genetically modified foods: an agenda for future research. Environ Health Perspect 2003; 111:1140–1141.
17. Workshop on transgenics. NTP Update 2003:3–4.
18. Jahnke GD, Choksi NY, Moore JA, Shelby MD. Thyroid toxicants: assessing reproductive health effects. Environ Health Perspect 2003; 112(3):363–368.
19. National Toxicology Program's Report of the Endocrine Disruptors Low-Dose Peer Review. Research Triangle Park, NC: National Toxicology Program, U.S. Department of Health and Human Services, Public Health Service, 2001.
20. Bucher JR. Update on National Toxicology Program (NTP) assays with genetically altered or "transgenic" mice. Environ Health Perspect 1998; 106(10):619–621.
21. Pritchard JB, French JE, Davis BJ, Haseman JK. The role of transgenic mouse models in carcinogen identification. Environ Health Perspect 2003; 111(4):444–454.
22. Haseman JK, Huff JE, Zeiger E, McConnell EE. Comparative results of 327 chemical carcinogenicity studies. Environ Health Perspect 1987; 74:229–235.
23. Huff J, Haseman J, Rall D. Scientific concepts, value, and significance of chemical carcinogenesis studies. Annu Rev Pharmacol Toxicol 1991; 31:621–652.
24. Bucher JR. The National Toxicology Program rodent bioassay. Ann NY Acad Sci 2002; 982:198–207.
25. Chhabra RS, Huff JE, Schwetz BS, Selkirk J. An overview of prechronic and chronic toxicity/carcinogenicity experimental study designs and criteria used by the National Toxicology Program. Environ Health Perspect 1990; 86:313–321.
26. http://ntp-server.niehs.nih.gov
27. Lucier GW. Mechanism-based toxicology in cancer risk assessment: implications for research, regulation, and legislation. Environ Health Perspect 1996; 104(1):84–88.
28. National Toxicology Program Fiscal Year 1984 Annual Plan. Research Triangle Park, NC: National Toxicology Program, U.S. Department of Health and Human Services, Public Health Service, 1984.

29. Levels of evidence of carcinogenicity used to describe evaluative conclusions for NTP long-term toxicology and carcinogenesis studies; request for comments. Federal Register 1986; 51(12):2579–2582.
30. http://ntp-server.niehs.nih.gov/htdocs/LT-studies/about-abstracts.html# CARCDEF.
31. Weisburger EK. History of the bioassay program of the National Cancer Institute. Prog Exp Tumor Res 1983; 26:187–201.
32. National Toxicology Program (NTP) Board of Scientific Counselors Technical Reports Review Subcommittee meeting; review of draft NTP technical reports. Federal Register 2003; 68(73):18,666–18,667.
33. http://ehp.niehs.nih.gov
34. http://www.iarc.int
35. Tomatis L. The IARC program on the evaluation of the carcinogenic risk of chemicals to man. Ann NY Acad Sci 1976; 271:396–409.
36. IARC. IARC Monographs on the Evaluation of Carcinogenic Risks to Humans. Overall Evaluations of Carcinogenicity: An Updating of *IARC Monographs* Volumes 1 to 42. Lyon, France: International Agency for Research on Cancer, World Health Organization, 1987.
37. Lists of IARC Evaluations. International Agency for Research on Cancer, World Health Organization, 2004. (Accessed March 28, 2004, at http://monographs.iarc.fr/monoeval/grlist.html)
38. http://www.epa.gov/iris/
39. What is IRIS? National Center for Environmental Assessment, Office of Research and Development, U.S. Environmental Protection Agency, 2004. (Accessed January 29, 2004, at http://www.epa.gov/iris/intro.htm)
40. Mills A, Foureman GL. US EPA's IRIS pilot program: establishing IRIS as a centralized, peer-reviewed data base with agency consensus. Toxicology 1998; 127(1–3):85–95.
41. FDA. Indirect food additives: adjuvants, production aids, and sanitizers. Federal Register 1998; 63(212):59,213–59,215.
42. FDA. Indirect food additives: paper and paperboard components. Federal Register 1999; 64(240):69,898–69,901.
43. FDA. Indirect food additives: polymers. Federal Register 2000; 65(14): 3384–3386.
44. FDA. Laxative drug products for over-the-counter human use. Federal Register 1999; 64(19):4535–4540.
45. Lewis CJ. FDA advises dietary supplement manufacturers to remove comfrey products from the market, 2001.
46. Hooper K, LaDou J, Rosenbaum JS, Book SA. Regulation of priority carcinogens and reproductive or developmental toxicants. Am J Ind Med 1992; 22(6):793–808.
47. Kizer KW, Warriner TE, Book SA. Sound science in the implementation of public policy: a case report on California's Proposition 65. JAMA 1988; 260(7):951–955.
48. The National Toxicology Program processes in relation to the authoritative bodies mechanism in Proposition 65. Office of Environmental Health Hazard Assessment, California Environmental Protection Agency, 2003. (Accessed March 28, 2004, at http://www.oehha.ca.gov/prop65/policy_procedure/ntptechrev.html)

49. http://www.oehha.org/prop65.html
50. National Toxicology Program Annual Plan for Fiscal Year 1995. Research Triangle Park, NC: National Toxicology Program, Department of Health and Human Services, Public Health Service, 1995.
51. Huff J. NTP Report on Carcinogens: history, concepts, procedure, processes. Eur J Oncol 1998; 3:343–355.
52. Report on Carcinogens. 10th ed. Research Triangle Park, NC: National Toxicology Program, U.S. Department of Health and Human Services, Public Health Service, 2002.
53. Barnard RC, Moolenaar RJ, Stevenson DE. IARC and HHS lists of carcinogens; regulatory use based on misunderstanding of the scope and purpose of the lists. Regul Toxicol Pharmacol 1989; 9(1):81–97.
54. OSHA. Occupational exposure to 1,3-butadiene. Federal Register 1996; 61:56,746–56,856.
55. OSHA. Occupational exposure to methylene chloride; final rule. Federal Register 1997; 62(7):1494–1543.
56. AFL-CIO et al., Petitioners and Plaintiffs v. George Deukmejian, et al., Respondent and Defendant. In: Superior Court of the State of California County of Sacramento, 1987.
57. Clarification of chemical listing of nickel and nickel compounds and request for comment on proposed listing of nickel compounds as known to cause cancer. Office of Environmental Health Hazard Assessment, California Department of Environmental Protection, 2004. (Accessed May 15, 2004, at http://www.oehha.ca.gov/prop65/prop65_list/0606NotNi.html)
58. Chemical listed effective May 7, 2004 as known to the State of California to cause cancer: nickel compounds. Office of Environmental Health Hazard Assessment, California Department of Environmental Protection, 2004. (Accessed May 15, 2004, at http://www.oehha.ca.gov/prop65/prop65_list/050704list.html)
59. Second Report Annual Report on Carcinogens. Research Triangle Park, NC: National Toxicology Program, U.S. Department of Health and Human Services, Public Health Service, 1981.
60. Report on Carcinogens. 9th ed. Research Triangle Park, NC: National Toxicology Program, U.S. Department of Health and Human Services, Public Health Service, 2000.
61. IARC. Some chemicals that cause renal or urinary bladder tumours in rodents, and some other substances. Saccharin and its Salts. IARC Monographs on the Evaluation of Carcinogenic Risks to Humans. Lyon, France: International Agency for Research on Cancer, 1999.
62. Chemicals delisted effective April 6, 2001 as known to the state to cause cancer. Office of Environmental Health Hazard Assessment, California Department of Environmental Protection, 2001. (Accessed March 28, 2004, at http://www.oehha.org/prop65/crnr_notices/chemicals_reconsideration/fdelistsacc.html)
63. http://cerhr.niehs.nih.gov/news/guidelines.html
64. NTP-CERHR Expert Panel Report on Di(2-ethylhexyl)phthalate. Research Triangle Park, NC: Center for the Evaluation of Risks to Human Reproduction, National Toxicology Program, U.S. Department of Health and Human Services, Public Health Service, 2000.

65. Expert Advisory Panel on DEHP in Medical Devices: Therapeutics Products Directorate Expert Advisory Panel on Di(2-ethylhexyl)phthalate in Medical Devices, Therapeutic Products Directorate, Health Canada, 2002.

66. CHAP. Report to the U.S. Consumer Product Safety Commission by the Chronic Hazard Advisory Panel on Diisononyl Phthalate (DINP). Bethesda, MD: U.S. Consumer Product Safety Commission Directorate for Health Sciences, 2001.

67. NTP-CERHR Expert Panel Report on Diisononyl Phthalate. Research Triangle Park, NC: Center for the Evaluation of Risks to Human Reproduction, National Toxicology Program, U.S. Department of Health and Human Services, Public Health Service, 2000.

68. NTP-CERHR Expert Panel Report on the Reproductive and Developmental Toxicity of 1-Bromopropane. Research Triangle Park, NC: Center for the Evaluation of Risks to Human Reproduction, National Toxicology Program, Department of Health and Human Services, U.S. Public Health Service, 2002.

69. EPA. Protection of stratospheric ozone: listing of substitutes for ozone-depleting substances—*n*-propyl bromide; proposed rule. Federal Register 2003; 68(106):33,284–33,316.

70. NTP-CERHR Expert Panel Report on the Reproductive and Developmental Toxicity of 2-Bromopropane. Research Triangle Park, NC: National Institute of Environmental Health Sciences, National Institutes of Health, 2002.

71. Meeting of the Developmental and Reproductive (DART) Identification Committee. Sacramento, CA: Office of Environmental Health Hazard Assessment, California Environmental Protection Agency, July 27, 1998.

72. OEHHA staff transparencies from meeting: Developmental and Reproductive Toxicant Identification Committee held on December 4, 2002. Office of Environmental Health Hazard Assessment, California Environmental Protection Agency, 2002. (Accessed March 28, 2004, at http://www.oehha.ca.gov/prop65/public_meetings/DART120402.html)

73. Summary of regulatory actions: Title 22. California Regulatory Notice Register 2003:332.

74. Chemicals under consideration for possible listing via the authoritative bodies mechanisms: request for relevant information. Office of Environmental Health Hazard Assessment, California Department of Environmental Protection, 2004. (Accessed May 28, 2004, at http://www.oehha.ca.gov/prop65/CRNR_notices/admin_listing/requests_info/dcallin21.html#get)

7

Toxicogenomics and Regulatory Science

Cynthia A. Afshari

Amgen Inc., Thousand Oaks, California, U.S.A.

INTRODUCTION

Since very early in the history of medicine, laypersons and medical professionals have pondered over the etiology of disease. Excessive exposure to exogenous agents such as drugs, water, food, or airborne agents led to the hypothesis that the basic chemical or contaminant properties of these agents, coupled with host factors and body burden, may be responsible for disease etiology or progression in some diseases. This hypothesis has fueled the perpetual refinement of the scientific discipline of toxicology along with the problem of optimal determination of risk.

In 2003, the world of science celebrated the 50th anniversary of the discovery of the structure of DNA by Watson and Crick. There was additional cause for celebration in 2003 with the benchmarking of the conclusion of a large, multinational, collaborative project to sequence the human genome, bringing to a closure the work to understand the DNA structure (1). However, with this new sequence information came the evolution of an infinite number of questions focused on how the information of the human genetic code translates into the individualized function of cells, organs, and organisms. Along with the new questions that scientists were now poised to address came a hope that this new genetic knowledge would serve as the backbone for novel discoveries that would shed light on the mystery of human disease, not only to cure it, but also to prevent it in future generations.

The hopes stemming from the new knowledge of the genome lie in the opportunity to develop a better understanding of disease, and the discovery of novel markers for staging disease and effects of chemicals and therapeutics. For this promise to be realized, there is a lot of work that needs to be carried on in the areas of marker discovery, and biomarker and technology validation. In addition, a framework for regulatory approval of new diagnostic products or registration of drugs and chemicals where genomics provides key supportive data needs to be put in place.

Two words that have come into more common use in the past decade and are descriptive of the integration of genomics science into the consideration of its impact of drugs and chemicals, are "pharmacogenomics" and "toxicogenomics." Pharmacogenomics is the discipline where investigators study the impact of drugs on the genome or how the state of an individual's genome may affect their response to a drug. "Pharmacogenomics" generally refers to the pharmacological effects of a drug or effects related to anticipated/intended use. In some cases, it refers to the genetic makeup of an individual with respect to the genetic coding of various polymorphisms that may affect cellular (i.e., enzymatic) activities leading to sensitivity or resistance to chemical exposure. When searching the public medical research literature database, one can find the first use of the term "pharmacogenomics" in 1961 (2,3). While there are only about 700 papers that used this term between 1961 and 2000, there have been over 2300 papers on this topic since 2000 showing the rapid development of this field over the past five years.

"Toxicogenomics" is a term that describes the functional interaction of the genome with agents that cause adverse effects in organisms. This is a new discipline with the first citation appearing in 1999 (4). Since then, there have been approximately 500 publications on toxicogenomics or closely related topics. For toxicologists who are focused on understanding the effects of exogenous agents on human health, there is hope that this new genomic information will bring enlightenment through new mechanistic readouts. The increase in mechanistic understanding of compound-organism interaction may ultimately lead to increased certainty in models of risk assessment.

Much of the technological applications and data analyses, in addition to the integration of genomics data into risk assessment, may be similar regardless of whether a study is characterized as being pharmacogenomic or toxicogenomic. There are some contextual differences in the sense of what is at stake in the two disciplines. For example, in pharmacogenomics, the science is driven toward ensuring that patients receive a drug that will allow a favorable response in ameliorating the symptoms of a disease; however, toxicogenomics is aimed at defining sensitive populations, or doses that may result in adverse effects, and therefore there is less tolerance of uncertainty in the application of this science in risk assessment designed to protect human health.

There has been some recent activity by multiple regulatory agencies on pharmacogenomics and/or toxicogenomics. Two of these agencies are the Federal Food and Drug Administration (FDA) and the Environmental Protection Agency (EPA). In this chapter the science of toxicogenomics will be broadly described in the context of the current activity within these two agencies.

This chapter will further briefly describe the technologies deployed in toxicogenomics studies and will use some case examples to illustrate this science. In addition, the current regulatory environment in toxicogenomics will be described.

MICROARRAYS IN TOXICOLOGY

Although the word "toxicogenomics" was not used until 1999, toxicologists were leveraging genomic information well before this. What was so groundbreaking in the emergence of toxicogenomics as a discipline was that the evolution of several new technologies that allowed scientists to progress from having to study the function/interaction of genes one at a time, to being able to consider the effects of the interaction of a compound and a biological system in the context of a whole genome, or thousands of genes simultaneously. One of the new technologies that led this new science was microarray technology.

First developed to study the effects of changing gene transcripts, cDNA microarrays allowed investigators to study a cell's or organ's response to a stimulus via a readout on effects of RNA message abundance across an entire genome (5–15). Subsequent to this, microarray chips have been developed to allow readouts of other biomolecules such as DNA copy number and proteins of all forms. The basic technologies associated with global analysis of ribonucleic acid (RNA) transcript, proteins, and metabolites are termed transcriptomics, proteomics, and metabonomics, respectively. Collectively, the field of toxicogenomics may unite all these types of technologies in an integrated fashion to completely understand the impact of an agent on an organism.

In the field of toxicogenomics, the DNA microarray chip is a commonly used tool to monitor the level of expression of a gene at the RNA level (5–7). The basic premise of transcript profiling is that the expression of a gene is monitored by the measurement of the hybridization of a cDNA synthesized from the transcript to a target sequence localized to a specific region on a chip. To measure this hybridization, RNA extracted from a biological sample of interest, is reverse transcribed into cDNA that ideally represents a quantitative copy of genes expressed at the time of sample collection. This cDNA is labeled with a "tracking" molecule such as a radioactive or a fluorescent nucleotide, or an affinity molecule like biotin and hybridized to the chip. Each microarray chip contains thousands of gene targets that can represent a partial or even a complete genome. In theory, each labeled cDNA molecule binds to its appropriate complementary target sequence on the array. Quantitative imaging allows measurement of the amount of labeled cDNA that hybridized to each target sequence, resulting in the identification and relative quantification of the genes expressed in the original biological sample.

Several varieties of the DNA microarray. including deposition or "spotted" cDNA, spotted oligomer, or synthesized oligomer chips are used frequently throughout the scientific community. The cDNA microarray is comprised of a collection of partial gene sequences that are "spotted" individually into precise locations within the DNA chip (5–12). Multiple genes within a conserved functional family may have a high degree of sequence similarity, especially in domains responsible for catalytic functions. To measure the gene of interest without cross-reactivity with other family members, usually sequences that contain a majority of 3' untranslated region are used to make the arrays. This allows researchers to take advantage of the sequence diversity in this region that may be gene-specific and not conserved among related family members. There are additional forms of DNA microarrays is based on deposition or on-chip synthesis of oligonucleotides (13–15). These chips contain short oligomers ranging from 25 to 80 bases as the target sequences. There is decreased sensitivity of binding to each short oligomer compared to that of the long cDNAs; however, this is compensated for by the "printing" of multiple oligos per gene, and the signal that represents binding is calculated from the collective signal.

The advent of microarrays has impacted expression analysis by allowing analysis and experiments to move from studying the expression of one gene in several days to studying hundreds of thousands of gene expressions in a single day. For the first time, investigators can relatively quickly measure the expression of a complete genome across a large number of environmental stimuli (16–19). As such, microarrays have led to the advances of information regarding cellular or organismal response to chemical exposures in many experiments worldwide; thus comprising a bulk of the public domain examples in the new discipline of toxicogenomics.

PROTEOMICS AND METABONOMICS

In addition to high-density transcript analysis using DNA chips, researchers are also able to examine the high content of proteins and metabolites in cell culture systems, target organs, or body fluids using a whole host of technologies that are collectively called as proteomics and metabonomics or metabolomics, respectively. The reader is referred to several detailed reviews of these technologies (18,19). By blending the knowledge of both the expression of proteins and transcripts, it may be possible to gain a broader perspective on the mechanism of action of toxicants because some functional pathway changes may be better reflected in mRNA changes and others in protein changes or modifications. One major advantage of proteomics and metabonomics studies is that measures may be made in body fluids such as serum, plasma, and urine thus allowing a noninvasive sampling of exposed individuals or animals over time. Metabonomics has indeed been shown to be a sensitive assay that reflects the toxicity in target organs by detecting it in urine (20,21).

CASE EXAMPLES

Scientists have conducted a number of proof-of-concept experiments to test the application of toxicogenomics toward several goals. First is towards the classification of compounds for either primary or secondary effect. Second is for the discovery of new safety biomarkers that are more sensitive and are early indicators of potential adverse effects. Third, toxicogenomics is being utilized to better understand the mechanism of action of compounds that will in turn lead to better understanding of cross-species and dose–response extrapolations and therefore improve the risk assessment process.

There are a number of case examples for toxicogenomics, where investigators have used microarrays to classify compounds by mechanism of action or by the effects that they elicit (22–26). Subsequently, these classifications led to the development of potential predictive models that have allowed the correct classification of animals exposed to unknown or blinded compounds segregated by primary mechanism of action or by the type of target organ tissue evoked (25–27).

To provide more detail on one of these case studies, Hamadeh et al. (24,25) developed gene expression profiles from the livers of rats that had been treated with compounds from two major classes of primary activity—peroxisome proliferators (i.e., Wyeth, Clofibrate, and Gemfibrozil) and a hepatic enzyme inducer, phenobarbital. These investigators built a classification model that allowed correct assignment of blinded compounds into each of the two classes, or to neither class. This proof-of-concept experiment was powerful in that it was an early demonstration that compound classification of unknown agents could be made against an existing set of gene expression profiles. However, this example also demonstrates the necessity for a database from which such classification models may be derived and the activity of unique compounds may not be understood with this approach. Investigators are hopeful that an approach where one builds up a set of data around classes of agents of interest or concern will permit future inferences about new compounds to be made.

Another hope of toxicogenomics classification lies in the ability to predict adverse outcome before it can be observed with histopathology. If found true, this could help build efficiency into the drug or chemical screening process. Steiner et al. (26) demonstrated the utility of a toxicogenomics database that was built to classify not around primary compound action, like the previous example, but around secondary effects such as types of hepatotoxicity, including steatosis and cholestasis. These investigators demonstrate that a compound's effects can be correctly predicted using comparison of the profile that it elicits in exposed tissue before the pathologist can observe the effects. Although this group has progressed in its prediction of compounds' potential to induce hepatoxicity it will not predict other types of toxicities in other target organs. Therefore, over time investigators will need to build similar models for additional disease endpoints. This work will take a large amount of resources and some significant time to achieve.

In the genomes that are normally studied by toxicologists, more than half the genes on most microarrays are not fully identified beyond the DNA sequence. Therefore these genes cannot be annotated in terms of mechanistic or pathway activity. This means that more than 50% of the data we are generating today may be uninterpretable. However, as genomic information increases and gene annotation improves, gradually new lessons will be gleaned from looking at previously derived microarray data or more powerful information will be deduced from new experiments. However, in spite of this limited annotation information, scientists are able to infer mechanistic information. In a study by Ueda et al. (28), the investigators used a constitutive androstane receptor (CAR) knockout model to understand the role of CAR in the evolution of liver's response to phenobarbital. Another interesting mechanistic study was aimed at the effects of arsenic in yeast (29). Yeast is a highly annotated genome, and in this study, the gene expression profiles could be mapped onto a very detailed pathway map that reflects gene–gene interactions, Cytoscape (30).

Finally, there are a few case examples where gene expression profiles have led to the elucidation of potential new safety biomarkers (31,32). In one of these studies, a new marker of gastrointestinal (GI) toxicity, adipsin, was elucidated from a very thorough and systematic approach to prioritize significant changes in gene expression data for those products that secreted proteins associated with GI toxicity. The emphasis on secreted proteins in their priority allowed further validation studies of the genomics marker to be conducted in noninvasive biological sampling (i.e., feces).

TOXICOGENOMICS AND THE REGULATORY ENVIRONMENT

Currently, microarrays are not routinely applied in regulatory settings, but there is increased interest on the part of basic and industry scientists as well as regulatory scientists to work toward this goal. Investigators have conducted proof-of-concept studies to standardize microarrays and their initial analyses in a regulatory setting (33–35). This technical validation is the first step to ensure a harmony between key stakeholders and the regulatory community in the interpretation of these data.

Chemical Industry

In 2002, EPA released an interim policy on the use of genomics in risk assessment (36). This interim policy succinctly stated that the EPA supported the use of genomics to understand the action of a compound in the environment or exposed individuals. The policy generally states that the EPA is in favor of the use of genomics endpoints to understand compound activity but that these data could not serve as the basis of regulatory decisions. While EPA was commended for reacting quickly in embracing toxicogenomics, this policy, however, lacks sufficient detail to provide a framework on how these data should be submitted and how the EPA regulatory review committee would handle them.

However, EPA followed this initial policy statement with the assembly of a Genomics Task Force that was charged with further examining the implications of genomics technologies on the risk assessment process. This group produced an additional document—"Potential Implications of Genomics for Regulatory and Risk Assessment Applications at the EPA" (37). The purpose of this document was to present exemplary applications and implications of genomics technologies into scenarios that could represent potential future circumstances under which such data might be presented to the EPA as part of a risk assessment paradigm. Four major areas were identified to be potentially impacted by genomics. They are: (i) prioritization of contaminants and contaminated sites, (ii) monitoring, (iii) reporting provisions, and (iv) risk assessment. The main areas that were identified for necessary research to support these scenarios were in the linking of genomics data to adverse outcome and in interpreting genomics information for risk and hazard assessment. In addition, it identified a need to hire scientists with expertise in this arena into the agency to formulate a strategy to extend training on the analysis and interpretation of these data to the EPA risk assessors. On the whole, the EPA is willing to be proactive in furthering the application of toxicogenomics into its decision-making process and to do so by working collaboratively with stakeholders.

Pharmaceutical Industry

In 2003, the FDA published the first draft of its guidance on pharmacogenomics and toxicogenomics to its regulated community (38). After some time, the FDA released a final version following revision driven by public comment. This guidance document, titled "Guidance for Industry: Pharmacogenomic Data Submissions" was released in March 2005 (39). The driving force behind the release of this document was the need to provide a regulatory framework for the submission of genomics data to the FDA in support of Investigational New Drugs, New Drug Applications, and Biologics License Applications with the hope that this would facilitate the application of genomics into nonclinical and clinical work of the pharmaceutical industry. In the document, "Critical Path to New Medicinal Products," the FDA administration suggests that the integration of these types of approaches into the drug development process is a necessity, to provide the public with novel and improved therapeutics that have increased efficacy and decreased adverse risk (40). The FDA's guidance document on pharmacogenomics has three main goals. These are to guide on when pharmacogenomic data should be submitted, what format and content should be included in the genomic submission, and how the data will affect the overall drug candidate review. The scope of the current guidance document excludes genomic data that is being used for characterization or quality control of biological product and metabonomic and proteomic technologies.

FDA recognizes that pharmaco- or toxicogenomics is still a new science and that most of the data coming forward may not be fully validated either in the technical or biological context (41). However, in its critical path document, the FDA indicates that to ultimately provide improved therapeutic benefit to the country's citizens, it needs to take the lead in driving new science and in the drug development process (40). Therefore, this genomics guidance document provides for a new submission scenario—the Voluntary Genomics Data Submission (VGDS) (39). The purpose of the VGDS is to allow sponsors to have the option of submitting genomic data to the FDA for review and to comment on where this analysis will not affect a regulatory decision. The VGDS is reviewed by a specialized group within the FDA, the cross-center Interdisciplinary Pharmacogenomic Review Group (IPRG), which reviews the data and then leads a discussion with the sponsor regarding the conclusions that the agency would draw from such data. In turn, the agency can learn about the structure, applications, and interpretations of these data in industry or on behalf of the stakeholder for the main purpose of joint education of both parties. The IPRG will work on the development of new policy in this arena, and when requested will serve as advisors to the review divisions on genomic data contained within submissions.

The FDA guidance document recommends submission of pharmacogenomic and toxicogenomic data to regulatory filings when the data of this type are used for decision making in the development of the proposed drug product. The document outlines that data from known valid biomarkers would be expected to be submitted to the agency, while that of probable or exploratory biomarkers may be submitted with regulatory applications or through the VGDS. The definition of the valid biomarker is important for investigators to understand. In the context of the FDA's view, a "valid biomarker" is defined as a marker that is measured in an analytical system with well-established performance characteristics and for which there is an established scientific framework or body of evidence that indicates significance of results. The FDA guidance document contains a glossary that clearly states their definition of these terms and also contains a number of decision trees that guide sponsors on the expectation regarding submission of their biomarker data in the context of the type of decision making and regulatory filing they are preparing (39).

CONCLUSIONS

The technologies that are used to support the discipline of toxicogenomics are rapidly maturing as scientists and technologists are gaining experience and collectively sharing best practices. The mechanistic interpretation of the data, however, is still challenged by incomplete genomic annotation and a lack of understanding of the biological meaning of the patterns derived from these data. However, scientists need to keep driving for the pieces of information that will fully elucidate these datasets. As scientists gain more confidence in the interpretation of the biological meaning of microarray/proteomic/metabonomic data, they will be successful in their quest to apply these data to understanding the complete aspects of interaction of chemicals on biological systems and thus more accurately inform the risk assessment process.

The complete biological meaning and context of such multivariate datasets will not be realized within a short window of time. Therefore, it is necessary for regulators and scientists to collaborate in exploring the meaning of these datasets. This collaboration requires "ground rules" that allow industry stakeholders to generate data with these technologies and develop case examples that demonstrate that these data can provide value toward the risk assessment process without the fear of unfounded impediment placed on the progression of the development of their potential products. In turn, these data need to be turned over to the regulatory agencies so that they can gain experience in determining how they will examine these data in the context of traditional regulatory submissions. Together, both industry and regulatory scientists need to synergize their perspective on the analysis and utilization of these datasets. The development of the current sets of policies from regulatory bodies of both the FDA and EPA have elicited a healthy environment for this collaboration and thus will hopefully ensure the promising impact of the genome era on better informing the risk assessment process.

REFERENCES

1. Lander ES, Linton LM, Birren B, et al. Initial sequencing and analysis of the human genome. Nature 2001; 409:860–921.
2. PubMed website: http://www.ncbi.nlm.nih.gov/entrez
3. Evans DA, Clarke CA. Pharmacogenetics. Br Med Bull 1961; 17:234–240.

4. Nuwaysir EF, Bittner M, Trent J, et al. Microarrays and toxicology: the advent of toxicogenomics. Mol Carcinog 1999; 24:153–159.
5. Schena M, Shalon D, Davis RW, et al. Quantitative monitoring of gene expression patterns with a complementary DNA microarray. Science 1995; 270: 467–470.
6. DeRisi J, Penland L, Brown PO, et al. Use of a cDNA microarray to analyze gene expression patterns in human cancer. Nat Genet 1996; 14:457–460.
7. Shalon D, Smith SJ, Brown PO. A DNA microarray system for analyzing complex DNA samples using two-color fluorescent probe hybridization. Genome Res 1996; 6:639–645.
8. Khan J, Saal LH, Bittner ML, et al. Expression profiling in cancer using cDNA microarrays. Electrophoresis 1999; 20:223–229.
9. Hegde P, Qi R, Abernathy K, et al. A concise guide to cDNA microarray analysis. Biotechniques 2000; 29:548–550, 552–556.
10. Spellman PT, Sherlock G, Zhang MQ, et al. Comprehensive identification of cell cycle-regulated genes of the yeast *Saccharomyces cerevisiae* by microarray hybridization. Mol Biol Cell 1998; 9:3273–3297.
11. Brown PO, Botstein D. Exploring the new world of the genome with DNA microarrays. Nat Genet 1999; 21:33–37.
12. Schena M, Shalon D, Heller R, et al. Parallel human genome analysis: microarray-based expression monitoring of 1000 genes. Proc Natl Acad Sci USA 1996; 93:10,614–10,619.
13. Lipshutz RJ, Fodor SP, Gingeras TR, et al. High density synthetic oligonucleotide arrays. Nat Genet 1999; 21:20–24.
14. Singh-Gasson S, Green RD, Yue Y, et al. Maskless fabrication of light-directed oligonucleotide microarrays using a digital micromirror array. Nat Biotechnol 1999; 17:974–978.
15. Hughes TR, Mao M, Jones AR, et al. Expression profiling using microarrays fabricated by an ink-jet oligonucleotide synthesizer. Nat Biotechnol 2001; 19:342–347.
16. Roberts CJ, Nelson B, Marton MJ, et al. Signaling and circuitry of multiple MAPK pathways revealed by a matrix of global gene expression profiles. Science 2000; 287:873–880.
17. Jelinsky SA, Estep P, Church GM, et al. Regulatory networks revealed by transcriptional profiling of damaged *Saccharomyces cerevisiae* cells: Rpn4 links base excision repair with proteasomes. Mol Cell Biol 2000; 20:8157–8167.
18. Liebler DC. Proteomic approaches to characterize protein modifications: new tools to study the effects of environmental exposures. Environ Health Perspect 2002; 110:3–9.
19. Merrick AB. Introduction to high-throughput protein expression. In: Hamadeh H, Afshari CA, eds. Toxicogenomics: Principles and Applications. Hoboken, New Jersey: John Wiley & Sons Inc., 2004:263–281.
20. Lindon JC, Holmes E, Nicholson JK. Metabonomics: systems biology in pharmaceutical research and development. Curr Opin Mol Ther 2004; 6:265–272.
21. London RE, Houck DR. Introduction to metabolomics and metabolic profiling. In: Hamadeh H, Afshari CA, eds. Toxicogenomics: Principles and Applications. Hoboken, New Jersey: John Wiley & Sons Inc., 2004:299–340.

22. Thomas RS, Rank DR, Penn SG, et al. Identification of toxicologically predictive gene sets using cDNA microarrays. Mol Pharmacol 2001; 60:1189–1194.
23. Waring JF, Gum R, Morfitt D, et al. Identifying toxic mechanisms using DNA microarrays: evidence that an experimental inhibitor of cell adhesion molecule expression signals through the aryl hydrocarbon nuclear receptor. Toxicology 2002; 181:537–550.
24. Hamadeh HK, Bushel PR, Jayadev S, et al. Gene expression analysis reveals chemical-specific profiles. Toxicol Sci 2002; 67:219–231.
25. Hamadeh HK, Bushel PR, Jayadev S, et al. Prediction of compound signature using high density gene expression profiling. Toxicol Sci 2002; 67:232–240.
26. Steiner G, Suter L, Boess F, et al. Discriminating different classes of toxicants by transcript profiling. Environ Health Perspect 2004; 112:1236–1248.
27. Thukral SK, Nordone PJ, Hu R, et al. Prediction of nephrotoxicants action and identification of candidate toxicity-related biomarkers. Toxicol Pathol 2005; 33:343–355.
28. Ueda A, Hamadeh HK, Webb HK, et al. Diverse roles of the nuclear orphan receptor CAR in regulating hepatic genes in response to phenobarbital. Mol Pharmacol 2002; 61:1–6.
29. Haugen AC, Kelley R, Collins JB, et al. Integrating phenotypic and expression profiles to map arsenic-response networks. Genome Biol 2004; 5:R95.
30. Shannon P, Markiel A, Ozier O, et al. Cytoscape: a software environment for integrated models of biomolecular interaction networks. Genome Res 2003; 13:2498–2504.
31. Amin RP, Vickers AE, Sistare F, et al. Identification of putative gene based markers of renal toxicity. Environ Health Perspect 2004; 112:465–479.
32. Searfoss GH, Jordan WH, Calligaro DO, et al. Adipsin, a biomarker of gastrointestinal toxicity mediated by a functional gamma-secretase inhibitor. J Biol Chem 2003; 278:46,107–46,116.
33. Irizarry RA, Warren D, Spencer F, et al. Multiple-laboratory comparison of microarray platforms. Nat Meth 2005; 2:345–350.
34. Larkin JE, Frank BC, Gavras H, et al. Independence and reproducibility across microarray platforms. Nat Meth 2005; 2:337–344.
35. Bammler T, Beyer RP, Bhattacharya S, et al. Standardizing global gene expression analysis between laboratories and across platforms. Nat Meth 2005; 2: 351–356.
36. Interim policy on genomics. U.S. Environmental Protection Agency, Science Policy Council, 2002:1–4; http://www.epa.gov/OSP/spc/genomics.pdf
37. Potential implications of genomics for regulatory and risk assessment applications at EPA. Genomics Task Force Workgroup, Science Policy Council, U.S. Environmental Protection Agency, 2004:1–48; http://www.epa.gov/OSA/genomics.htm
38. Guidance for Industry: Pharmacogenomic Data Submissions, Draft guidance. U.S. Food and Drug Administration, 2003; http://www.fda.gov/OHRMS/DOCKETS/98fr/2003d-0497-gd10001.pdf
39. Guidance for Industry: Pharmacogenomic Data Submissions, U.S. Food and Drug Administration, 2005:1–25; http://www.fda.gov/cder/guidance/6400fnl.pdf

40. Challenge and Opportunity on the Critical Path to New Medicinal Products, US Food and Drug Administration, March, 2004; http://www.fda.gov/oc/initiatives/criticalpath/whitepaper.html.
41. Lesko LJ, Woodcock J. Translation of pharmacogenomics and pharmacogenetics: a regulatory perspective. Nat Rev Drug Disc 2004; 3:763–769.

8

Physiologically Based Pharmacokinetics

William L. Roth

*Office of Food Additive Safety, U.S. Food and
Drug Administration, College Park, Maryland, U.S.A.*

INTRODUCTION: RELATIONSHIP OF PHARMACOKINETICS
TO TOXICOLOGY

Most of the disciplines that contribute to the field of toxicology determine
whether a chemical has an adverse effect on an organism, and characterize
any effects observed, sometimes to the molecular level. Pharmacokinetics is
concerned with predicting the concentration of a toxic or pharmacologically
active substance in the blood and tissues of the exposed/treated animal.

If other toxicological studies have determined a relationship between
the whole body dose and an effect, or between the concentration at a target
tissue and the effect, the dose–response model can be combined with a pharmacokinetic model to give a model which can predict the intensity of effect
at any dose. The combined kinetic/effect model is called a pharmacodynamic model. If the effect at the "target" is the same between two species
with different metabolic characteristics, a good pharmacodynamic model
should allow one to predict the difference in effect between species for a
particular dose, or determine what doses give equivalent responses.

The relationship between pharmacokinetic studies and toxicological
effect studies is illustrated in the information flow diagram that follows.

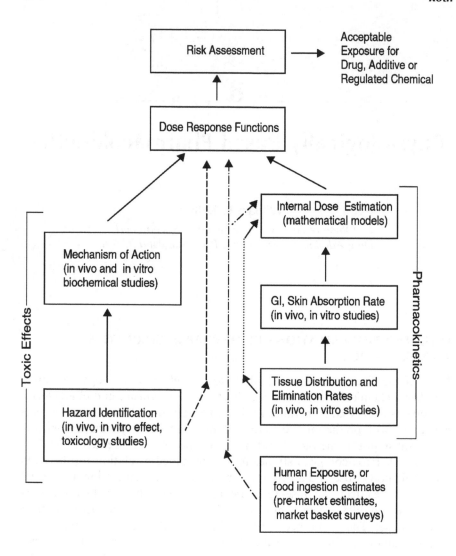

This chapter describes different approaches to building dose–response models which integrate anatomical, physiological, and biochemical properties of animals with the chemical properties and biological effects of regulated chemicals.

The study of pharmacokinetics has been made largely possible as a result of the availability and introduction of radioisotopes and stable isotopic tracers into medical science, beginning in the late 1940s.

"Pharmacokinetics" was coined as a term to describe the mathematical analysis of data generated by experimental measurement of the (*i*) absorption of chemicals into the blood, (*ii*) distribution from the blood into tissues, (*iii*) metabolism (biochemical transformation) of the parent chemical into any metabolites which a particular animal might be capable of producing from it, and (*iv*) elimination from the body of the parent chemical and its metabolites. A typical absorption, distribution, metabolism, and elimination (ADME) experiment would generally involve the injection of a group of animals with a radioactive tracer (chemical incorporating ^{3}H, ^{14}C, ^{32}P, ^{35}S, etc.) followed by blood sampling and sacrifice for collection of tissues at carefully planned and measured time points postinjection (Fig. 1). Blood and tissue samples would then be radiochemically analyzed for the tracer and any metabolites. The ADME data resulting from such experiments would constitute sets of data available for pharmacokinetic analysis (1).

Pharmacokinetic analysis is generally a complex process because of the complex nature of the biological systems under study. Exposure to chemicals generally occurs via oral, inhalation, and dermal routes, or portals of entry (POE), each of which incorporates physiological functions that dynamically limit the absorption of some chemicals and enhance the absorption of others. Effects of these route POE specific processes on absorption are generally incorporated by adding models of the POEs as input functions to the basic pharmacokinetic models. The physiological and anatomic characteristics of each of these POEs are described in greater detail in other

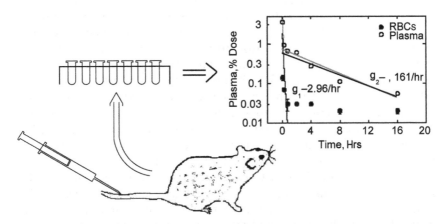

Figure 1 Typical ADME experiment involves IV injection of a labeled chemical (tracer) into test animals followed by collection of blood samples (and tissues) at carefully selected times postinjection. IV dosing time points are selected to ensure that distribution (into tissues) and metabolism/elimination phases (D and E) for the chemical are characterized. Oral dosing time points are selected to ensure characterization of absorption and metabolism/elimination phases (A and E). *Abbreviations*: ADME, absorption, distribution, metabolism, and elimination; IV, intravenous.

references (2,3). Distribution of the systemically absorbed chemical from the site of absorption into tissues can also become mathematically complex, depending on tissue/protein binding and metabolic pathways associated with the chemical. Some of these factors will be discussed in sections which follow.

Pharmacokinetic analysis, as practiced from the late 1950s to 1970s, involved the fitting of data sets (generally concentrations of tracer in blood, feces, and urine) to sets of one, two, or three differential equations which described the ADME processes thought to be occurring in one, two, or three idealized "compartments" (4). The volumes of these compartments did not generally correspond to the known blood volume or specific tissue volumes. These "classical" models generally incorporated volume in the form of a "volume of distribution" (V_d) which is the intercept of the distribution phase curve obtained in the curve fitting process (Fig. 2). Other parameters of classical models such as intercompartmental transport coefficients (k_{ij}) and the systemic elimination rate coefficient (k_e) are likewise obtained by curve fitting and additional model-dependent calculations. Algebraic solutions are generally available for all models composed of fewer than four simultaneous differential equations (5,6).

Several monographs are available which focus on the theory, mathematics, and construction of such classical models (4,6,7). The focus of this chapter is on the relationships between classical and physiologically based pharmacokinetic (PBPK) models, followed by examples of models used for purposes of regulating industrial chemicals, drugs, and food additives.

Over the past 20 years, "classical" pharmacokinetic models have been increasingly replaced by PBPK (8). PBPK models are designed to incorporate more realism, known anatomic and physiological characteristics of the POE, and physiological flow limitations for each test species in the model structure and the behavior. Like classical models, PBPK models are composed of sets of differential equations, incorporating the characteristics and limitations of the physiological parameters described above, but they generally require more than four simultaneous equations, describing more than four compartments (or volumes). As a consequence, PBPK models must generally be constructed and solved in several steps, including the curve fitting steps used to determine parameter values for classical models, followed by numerical integration and optimization, which are simply methods of accomplishing curve fitting for more complex systems.

Regulatory applications of pharmacokinetics have included the use of simple, one, two, or three-compartment models, primarily for the estimation of the parameters V_d and k_e by curve fitting. More complex PBPK models have been applied to problems which are more complex in terms of routes of exposure and modes of action (mechanism) as discussed in this chapter.

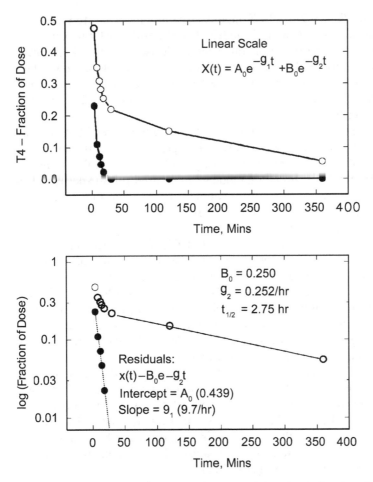

Figure 2 Curve fitting: fitting sum of exponentials to IV dosing data. Upper figure is a linear scale plot showing observed plasma concentrations (○) and a function to be fitted ($X(t) = \ldots$). The lower, logarithmic plot shows the calculated parameters for the elimination phase (B_0, g_2) and residual values (●) from which distribution phase parameters (A_0, g_1) can be calculated. *Abbreviation*: IV, intravenous.

RELATIONSHIP BETWEEN CLASSICAL PHARMACOKINETIC MODELS AND PHYSIOLOGICALLY BASED MODELS

Most parameters which are incorporated into classical and physiologically based models assume that the processes they are used to describe are of first order, i.e., directly proportional to the concentration of the chemical in a particular compartment. Parametric differences between classical and PBPK models are as follows:

a. *Blood flows* (Q_js). PBPK models generally incorporate blood and
    excreta flows. Classical models describe transport between com-
    partments via transport rate coefficients [see (d)] and do not gen-
    erally impose physiological limitations. A set of blood flow values
    for humans and traditional test animal species have been com-
    piled by Brown et al. (9), and are now considered to be well
    established.
b. *Tissue volumes* (V_js). Volumes relative to body weight (BW) may
    differ with species and change with age. Good growth models are
    available for the rat, mouse, dog, and human (Tables 1 and 2).
    Less accurate allometric models can be used for all species. Clas-
    sical models incorporate volumes of distribution (V_ds), not
    plasma or tissue volumes. PBPK models include measured
    plasma and tissue volumes.
c. *Partition coefficients* (R_{jp}s $= C_j/C_p$). Classical models rarely incor-
    porate R_{jp}s, which would be used with the assumption that a
    steady-state or rapid equilibrium is achieved. Partition coefficients
    add no dynamic information about transport. PBPK models invari-
    ably incorporate R_{jp}s, but rarely incorporate dynamic information
    associated with intercompartmental transport coefficients (k_{ij}s).
d. *Transport rate coefficients* (k_{ij}s). Classical models incorporate
    dynamic information as first-order transport rate constants (k_{ij}s).
    Physiological models almost always incorporate R_{jp}s, and therefore
    assume that rapid mixing/equilibrium conditions prevail and that
    chemical transport is limited by blood flow, rather than barriers.
e. *POEs.* PBPK models generally include mathematically detailed
    descriptions of POE, metabolism, and tissue distribution pro-
    cesses. The distribution models are often limited by the use of
    R_{jp}s as mentioned in (c) and (d).

Figure 3 shows the parameters and features of a three-compartment
classical model and its equivalent physiologically based model.

Differential equations which might be used to describe the models in
Figure 3 are given below.

Mammillary Models

$$\frac{dX_p}{dt} = -k_{j1}X_p + k_{1j}X_j \quad \frac{dX_j}{dt} = k_{j1}X_p - k_{1j}X_j - \cdots \tag{1}$$

Mammillary models generally incorporate first-order intercompartmental transport coefficients. The tissue–plasma transport rate constants
k_{j1} and k_{1j} have units of 1/unit time, which yield units for dX_p/dt of
mass/unit time because both constants are multiplied by the mass in blood

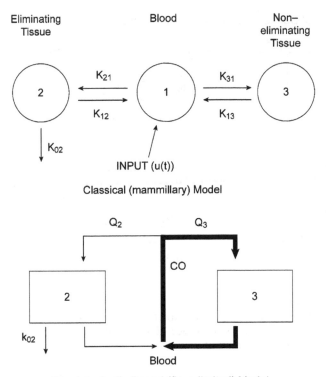

Classical (mammillary) Model

Physiologically Based (flow–limited) Model

Figure 3 Parameters and features of a three-compartment classical model and its equivalent physiologically based model. The k_{ij} are first-order intercompartmental transport coefficients (1/min), Q_i are blood flows (mL/min), and CO is the cardiac output (mL/min). Classical models with a "central" compartment are also referred to as "mammillary" models for reasons described by Rescigno and Segre (22).

or tissue. The first subscript of the k_{ij} represents the destination compartment, the second the origin. In this case, compartment 1, the "central" compartment, represents blood plasma. A metabolism term (a k_e-dependent term, say, $k_e X_j$) could be added to the tissue equation, if it represented metabolizing tissues such as the liver or the kidney.

PBPK Models

Physiologically based models always incorporate intercompartmental physiological flows, Q_js (volume/unit time), which limit the amount of chemical delivered to a tissue, especially for rapidly transported substances such as Na^+, O_2, ethanol, etc. Concentration changes in the blood between the arterial (C_p) and the venous side (C_{vj}) depend on partitioning and transport barriers (Fig. 4). The most common form (no diffusion barrier) of PBPK

Transport of Solute Between Tissue and Blood

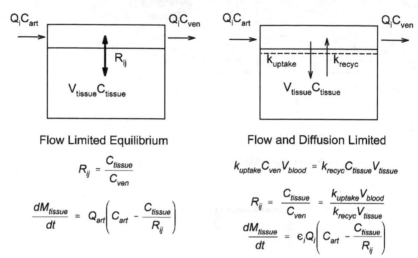

Figure 4 Comparison of flow-limited (equilibrium) transport model equations to flow and diffusion-limited model equations. The extraction efficiency (ε_i) is proportional to the permeability, blood flow, and surface area available for absorption within the organ/tissue for which it is derived.

equations are

$$\frac{dX_p}{dt} = -Q_j(C_p - C_{vj}) \quad \frac{dX_j}{dt} = Q_j(C_p - C_{vj}) - \cdots \tag{2}$$

The following definitions are generally assumed, but may seem to conflict:

$$C_{vj} = C_j/R_{jp} \quad C_p = X_p/V_p \quad \text{(units = mass/volume)}$$

$$C_j = X_j/V_j \quad R_{jp} = C_j/C_{vj} \quad \text{(partition coefficient)}$$

PBPK models assume that there is always a small difference between arterial plasma (C_p) and venous (C_{vj}) concentrations. With a diffusion barrier incorporated in the form of an extraction efficiency (ε_j), Equation 2 becomes:

$$\frac{dX_p}{dt} = -\varepsilon Q_j(C_p - C_j/R_{jp}) \quad \frac{dX_j}{dt} = \varepsilon Q_j(C_p - C_j/R_{jp}) - \cdots \tag{3}$$

The parameter ε_j is unitless, and is always less than 1. It represents a sort of probability that a molecule crosses the barrier between the compartments during a single pass (23).

If the terms of the physiological model [Eq. 3] are equated with the terms of the mammillary model [Eq. 1] the following relationships hold for in vivo study results:

$$\varepsilon_j Q_j C_p = k_{j1} X_p = k_{j1} C_p V_p \rightarrow \varepsilon_j = k_{j1} V_p / Q_j \quad \text{(extraction)} \tag{4}$$

$$\varepsilon_j Q_j C_j / R_{jp} = k_{1j} X_j = k_{1j} C_j V_j \rightarrow k_{1j} = \varepsilon_j (Q_j / V_j R_{jp}) \quad \text{(recycling)} \tag{5}$$

and

$$k_{1j} = k_{j1} V_p / V_j R_{1j} \rightarrow R_{jp} = k_{j1} V_p / k_{1j} V_j \quad \text{(partitioning)} \tag{6}$$

The tissue uptake parameter k_{j1} is obtained from the distribution phase of a plasma concentration curve (tissue concentration, if available) by fitting parameters to data from a classical intravenous (IV) injection experiment. The recycling parameter k_{1j} is obtained via nonlinear optimization, from an initial value estimated from the partition coefficient.

The Physical Meaning of a Partition Coefficient

Partition coefficients are a convenient way of incorporating the relatively constant ratios between blood and tissue concentrations, found at steady state, into pharmacokinetic models. However, partition coefficients should be recognized as approximations to complex binding and phase (lipid: aqueous) equilibria which may be resolved with carefully chosen in vitro studies on blood and tissue components (1,24).

As an example, let us simulate the value of a tissue: plasma "partition coefficient" of a chemical which binds to albumin. If total albumin concentrations ($P_t = [P] + [PL]$) are set to give plasma $P_t = 450 \ \mu M$, tissue albumin $P_t = 250 \ \mu M$, and a dissociation coefficient $K_d = [P][L]/[PL]$ (for both compartments), the relationships shown in Figure 5 are obtained. Note the rapid change in the "partition coefficient" (tissue: blood ratio) of the hypothetical chemical as the tissue binding sites become saturated. Real tissues may have multiple binding species, and more complex tissue:plasma binding curves than illustrated here.

Effect of Metabolism on Partitioning

Rapid elimination or metabolism in a compartment (say liver) may prevent that compartment from ever approaching equilibrium. Computation of partition coefficients is usually based on the assumption that at some time point as t increases, the rate of change of plasma drug mass approaches zero ($dX_p/dt \rightarrow 0$), so that rates of recycling of tissue drug to plasma, and plasma to tissue are approximately equal:

$$Q_j C_p \simeq Q_j C_j / R_{jp}$$

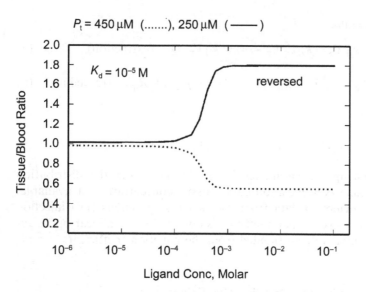

$P_t = 450\,\mu M$ (........), $250\,\mu M$ (——)

Figure 5 Ratio of concentrations (partition coefficient, R_{jp}) between hypothetical tissue and plasma compartments as determined by independent concentrations of a binding protein (P_t) present in plasma at $450\,\mu M$, and in tissue at $250\,\mu M$, with an equilibrium binding coefficient $K_d = 10^{-5}\,M$. If the plasma and tissue concentrations are reversed, the upper plot is obtained in the simulation. Inclusion of additional binding species or hydrophobic phases would increase the complexity.

In the absence of metabolism, the partition coefficient (R_{jp}) is the ratio of uptake to recycling rates [Eq. 6].

If metabolism is rapid ($\varepsilon_j Q_j C_p \ll k_e C_j V_j$), then the tissue concentration is determined primarily by the uptake rate, as mass cannot be removed faster than it appears. The partition coefficient can be obtained from concentrations in the "*j*" compartment mass balance at steady state:

$$\frac{dX_j}{dt} = k_{j1} C_p V_p - (k_{1j} + k_e) C_j V_j = 0 \qquad (7)$$

$$R_{jp} = C_j / C_p = k_{j1} V_p / (k_{1j} + k_e) V_j \qquad (8)$$

In such a situation, the partition coefficient remains constant, but is dependent on the metabolism coefficient (k_e). If $k_{1j} \ll k_e$ partitioning becomes approximately the ratio of uptake to metabolism: $R_{jp} = k_{j1} V_p / k_e V_j$ [compare to Eq. 6].

Pharmacokinetic models that incorporate metabolism frequently use Michaelis–Menten type expressions so that saturation of metabolic capacity can be modeled, with the enzyme reaction velocity v_j substituted for $k_e X_j$:

$$v_j = \frac{K_f C_j V_j}{K_m + C_j}$$

Note that the specific activity or capacity of the enzyme (K_f, mol/min g) is multiplied by the tissue volume (V_j, g/tissue) to replace the traditional maximal velocity (V_{max}) used to describe test-tube experiments. If we substitute v_j for k_e in differential Equation 7, the partition coefficient takes on different values at low and high tissue concentrations, assuming steady state:

$$\frac{dX_j}{dt} = k_{j1} C_p V_p - \left[k_{1j} + \frac{K_f}{K_m + C_j} \right] C_j V_j$$

$$R_{jp} = C_j / C_p = \frac{k_{j1} V_p / V_j}{k_{1j} + K_f / (K_m + C_j)}$$

(9)

Detailed descriptions of experiments and models employed to estimate partition coefficients in the presence of drug metabolism and/or elimination are provided in papers of Chen and Gross (25) and Gallo et al. (26).

ANATOMIC MODELS AND PHYSIOLOGICAL SCALING

Body weight (BW), tissue, and organ volumes (V_j) differ dramatically between animal species. These differences must be incorporated into PBPK models whenever an attempt is made to extrapolate pharmacological and/or toxicological effects or the underlying differences in pharmacokinetics between species. Extrapolation between short-term results (1–2 days) may be accomplished using standardized tables (9) of BW, V_j, and blood flows (Q_j).

Growth and Anatomical Models

When longer-term effects or chronic studies are to be simulated, the pharmacokinetic models used for extrapolation must account for continuous changes in food and water consumption, growth, altered physiological status, and possibly effects of aging if an accurate extrapolation is required. Increases in BW and tissue volume during the course of an experiment may result in the dilution of a bioaccumulative chemical giving the appearance of a faster elimination rate. Weight loss may result in increased concentrations of a slowly eliminated chemical, giving the appearance of very slow or negative elimination (27). An important example of such a situation is evident during the development of the fetus. Different components of the conceptus (placenta/fetus/amniotic fluid system) grow at very different rates during gestation, and the fetus may increase in volume by a factor of 10 within a 48-hour period (28). This rapid change in volume can disguise changes in the mass balance because of the rapid dilution that occurs with

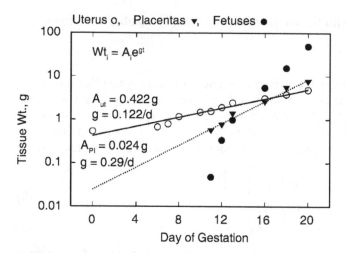

Figure 6 Growth of rat conceptus compartments and exponential curves fitted to the data. The parameter A_i is the intercept (weight) at gestation day 0, and parameter g is the first-order increase between time points. Fetal growth is best fitted by a Gompertz function, which is of the form $W(t) = W(0) \exp\{(\beta/\alpha) [1 - e^{-\alpha t}]\}$, where $W(0)$ is the starting fetal/embryonic weight, β is the initial growth rate, and α is the deceleration of growth from initial rate β. *Source*: From Ref. 28.

the rapid increase in fetal volume during such gestational periods. An example of the use of conceptus growth models within a pharmacokinetic analysis is shown in Figures 6 and 7. More detailed examples of fetal pharmacokinetic models may be found in Refs. 29 and 30.

Many species-specific growth models have been developed for incorporation into such analyses and kinetic models (29,31). Tables 1 and 2 provide some parameters and growth functions which have not been compiled in sources such as Brown et al. (9). Normal growth of male rats and their major organs are illustrated in Figure 8.

Physiology at the POE

Regulatory applications of PBPK models are generally concerned with one or more specific exposure scenarios. Two of the most important scenarios are oral dosing (chemicals in food and drugs) and inhalation exposure (occupational and environmental air). Contributions of the physiology of these POEs to chemical absorption and elimination are discussed below.

Gastrointestinal Clearance vs. Absorption

The accuracy of models of gastrointestinal (GI) absorption is determined in large part by the volume, length, fluid flows, and relative absorption properties

Table 1 Whole Body Growth Parameters $BW(t) = W_0 \exp\left[\frac{\beta_1}{\alpha_1}(1 - e^{-\alpha_1 t}) + \frac{\beta_2}{\alpha_2}(1 - e^{-\alpha_2 t})\right]$

Species/strain	Growth velocity and deceleration (1/day)				Data source(s)
	β_1	α_1	β_2	α_2	
Mouse					
		$W_0 = 1.48–0.5$ g (M), 1.38–0.4 g (F)			
C57 (male)	0.069	0.048 (6.428 = W_{14})	—	—	(10)
C57 (female)	0.067	0.053 (6.175 = W_{14})	—	—	(11)
DBA (male)	0.065	0.045 (7.273 = W_{14})	—	—	
DBA (female)	0.061	0.054 (7.406 = W_{14})	—	—	
		$W_0 = 1.42–0.4$ g (M), 1.39–0.4 g (F)			
CF-1 (male)	0.204	0.0685	—	—	(12)
CF-1 (female)	0.204	0.0735	—	—	(13)
B6C3F1 (male)	0.204	0.0725	—	—	(14)
B6C3F1 (female)	0.204	0.0780	—	—	(13)
Rat					
		$W_0 = 6.24–0.9$ g (M), 5.96–0.7 g (F)			
Sprague–Dawley male	0.142	0.03335	0.474E-03	0.142E-03	(15)
Sprague–Dawley female	0.141	0.0407	2.570E-03	4.430E-03	(11)
F-344 (male)	0.0914	0.0233	1.10E-03	2.68E-03	(14)
F-344 (female)	0.0914	0.0233	3.28E-03	4.00E-03	
Dog (beagle)					
		$W_0 = 285–55$ g			
Male	0.063	0.0181	—	—	(16)
Mini-Pigs (Hormel)					
		$W_0 = 0.591–0.139$			
Male	0.0760	0.015	176.6E-06	459.1E-06	(17)
Female	0.0611	0.012	146.6E-06	353.2E-06	(17)

Two component Gompertz growth (BW) equation from Ref. 9. Parameters β_i are initial velocities, α_i are deceleration coefficients for growth phases 1 and 2 (if present). The symbols W_0 and W_{14} represent BW at birth (0) and 14 days of age, respectively.
Abbreviation: BW, body weight.

of different segments of the GI tract. These anatomical and physiological properties are not necessarily scalable (with any accuracy) between species (32,33). Although the absolute lengths and volumes of intestinal segments are quite different, the cecum is probably has the most variable function

(A)

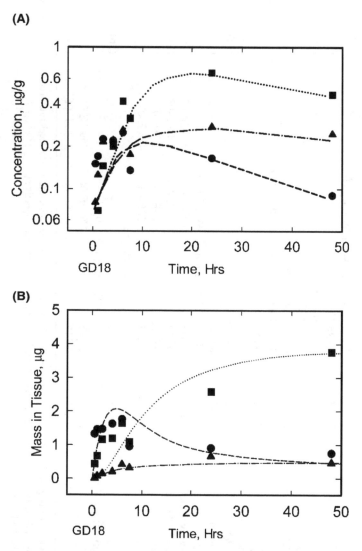

(B)

Figure 7 Mathematical transformation of concentrations of trichlorosucrose (TGS) in conceptus compartments into mass of TGS in placenta (▲), fetus (●), and amniotic fluid (■) compartments via anatomical growth models of the rat conceptus. The lines in the lower figure are physiologically based pharmacokinetic (PBPK) model simulations based on parameters from this data.

Table 2 Rat Organ Weight vs. Age and BW[a]

Organ/volume	Fitted equation (age in days)	Source
Blood	$0.0935 \, (BW^{0.9})$	c
Brain	$0.569 \log (BW - 0.87) + 0.554$	b
Heart	$0.0026 \, (BW + 14) + 0.249 \log(BW + 14) - 0.336$	b
Total fat	$BW(0.04 + 0.15(1 - \exp(-0.0055 \, age)))$	d
Liver	$0.0303(BW + 5) + 3.34 \log(BW + 5) - 3.896$	b
Lung	$0.00471(BW + 2) + 0.122 \log(BW + 2) - 0.056$	b
Kidney	$0.00718(BW - 3) + 0.132 \log(BW - 3) - 0.009$	b
Spleen	$0.00245 \, BW + 0.0301 \log(BW) - 0.025$	b
GI tract	$0.0245 \, BW + 4.72 \log(BW + 7) - 5.753$	b
Stomach	$0.0024 \, BW + 0.631 \log(BW) - 0.713$	b
Derived from other formulas		
Intestines	GI tract − stomach	c
Skin + muscle	BW − 3 organ weights − 0.7606 GI tract (contents)	c
Muscle	0.85 (skin + muscle)	c

[a]BW derived from experimental data or from Table 1 (equation and parameters).
[b]Equations from Ref. 18.
[c]Equations from Ref. 19.
[d]Ref. 19—data derived from Refs. 20 and 21.
Abbreviations: BW, body weight; GI, gastrointestinal.

between the species listed. In humans the cecum constitutes only 4% of the length of the large intestine, and is a small corner at the junction of the small and large intestines. Dogs have little or no cecal function. In the rat, the cecum constitutes 25% of the length of the large intestine, and is shaped like a bladder with a small inlet and outlet (34). Species differences in the length and function of intestinal segments (duodenum, jejunum, cecum, and colon) determine the extent and rapidity of absorption of different classes of chemicals from these segments. The cecum and large intestine of most animals contain a normal bacterial population which contributes significant mass to the lumen contents and may metabolize a drug prior to absorption or act on its metabolites on elimination. Such bacterial metabolism may dramatically affect drug bioavailability.

Stomach emptying rates are among the most important parameters in developing accurate GI absorption models. Many if not most absorbable substances are absorbed through the intestines, to which food and fluids (chyme) are metered by the stomach as it empties. The rate at which the stomach discharges its contents depends on the contents of the chyme (liquids and solids), characteristics (acidity, carbohydrate, fat content, etc.), and the volume contained in the stomach. Intestinal detection of the character of the food (glucose, digestible fat, and amino acid content) reaching it from the stomach results in feedback which can change the rate of stomach emptying by as much as a factor of 6 (35,36). Analysis of such results using models

Growth of Rat Tissues

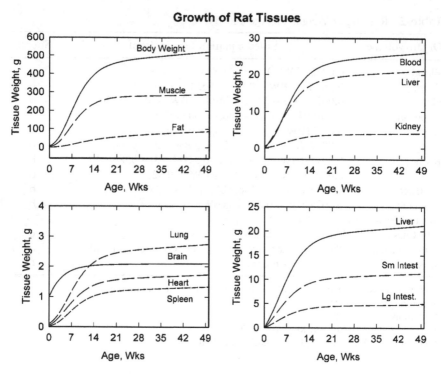

Figure 8 Growth of male Sprague–Dawley rats and major organs, as simulated by fitted equations and parameters listed in Tables 1 and 2. Growth of organs in the mouse and dog have also been fitted. *Source*: From Refs. 16, 18, and 19.

described later in this chapter shows that GI transit and physical properties of the food matrix can have as great an influence on the absorption of chemicals as the properties of the chemical itself.

In general terms, stomach emptying can be described as a first-order process, with an emptying coefficient (k_{empty}) which is dependent on the characteristics of food and liquid in the stomach:

$$\frac{dX_{sto}}{dt} = U(t) - k_{empty}X_{sto} \tag{10}$$

In Equation 10, X_{sto} is the mass of food contained in the stomach, and $U(t)$ is the rate of input into the stomach. Mixing of material with existing contents generally results in the maintenance of a first-order emptying condition. Several authors (37,38) studied gastric emptying in human subjects and found strong correlations between emptying rates, time to peak plasma levels of drug, and maximum plasma concentrations. A theoretical example of the within-species change in gastric emptying and its effects on plasma pharmacokinetics is shown in Figure 9. Additional examples

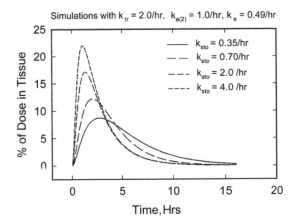

Figure 9 Change in time of appearance of $C_{\max}$ and fraction of dose in whole body with changing stomach emptying rate coefficient (k_{sto}), fixed intestinal transport coefficient (k_{tr}), and fixed systemic elimination rate (k_{e}). The fastest rate is similar to what would be expected for a fasted subject given a sugar solution orally, while the slowest rate is approaching the rate of emptying of a fatty meal and subsequent triglyceride absorption (Table 2).

are also described in this chapter. GI parameters for rats, dogs, and humans are listed in Table 3. A more detailed review of GI transit, drug absorption, and factors affecting transit rates has been published by Dressman et al. (39).

Determination of rates of GI absorption and intestinal metabolism of chemicals are dependent on GI clearance, which depends on food and water consumption, as mentioned above. Absorption from any particular segment is generally described in PBPK models by a first-order absorption coefficient ($K_{\text{a}(i)}$) or by a permeability coefficient (P_{eff}):

$$\frac{\mathrm{d}X_i}{\mathrm{d}t} = k_{\text{tr}(i-1)}X_{i-1} - (K_{\text{a}(i)} + k_{\text{tr}(i)})X_i \tag{11}$$

Here the parameters $k_{\text{tr}(i)}$ denote physiological transport coefficients (Table 3) of succeeding GI segments. Models may contain a single or multiple absorption sites.

A typical ADME study is not concerned with rates ($K_{\text{a}(i)}$) or sites ("i") of intestinal absorption, which determine the shape and peak concentrations of the plasma concentration curve. Generic ADME studies are more concerned with total systemic absorption or bioavailability (F). The fraction F can be estimated either from the ratio of IV to oral dose areas under the plasma concentration curve ($F = \text{AUC}_{\text{oral}}/\text{AUC}_{\text{IV}}$) or from the total residue in tissues and chemical residues eliminated in excreta. When computing a

Table 3 GI Tissue Weights, Volumes, and Emptying Coefficients

Segment/species	Adult tissue wt. (% BW)[a]	Lumen vol. (% BW)	Gastric/bile secretion	k_{empty} (1/hr)[b]	Comments
Rat (Sprague–Dawley)					
Stomach	0.50 ± 0.06	(1–3)	?	$1.15 \geq 0.33$	fasted → ad libitum fed
Small intestine	1.7 ± 0.5	(0.4)	65 µL/min/kg	0.63	($t_{1/2} = 1.1$ hr)
Cecum	0.27	(1.2)		$0.10 \geq 0.20$	($t_{1/2} = 6.9$ hr)
Large intestine	0.34	(1.2)		(7+ hr τ) 0.04	Stomach → rectum = 15 hr
Dog (beagle)					
Stomach [wt. $= 11.9(\mathbf{BW})^{1.23}$]	0.8 ± 0.15	(0–6)	?	$1.4 \geq 6.0$	w/55 mM glucose → saline
				0.85–1.2	w/500 mM glucose
				0.35	w/meat cubes
Small intestine	2.2 ± 0.7		6 µL/min/kg	(12–16 hr τ)	Total GI system
Large intestine [total $= 64(\mathbf{BW})^{1.04}$]	0.7 ± 0.05				
Human					
Stomach	0.20	1300 mL (1.8)	30–700 mL/hr	1.1–4.6	330–1250 mL
				2.2–0.9	sucrose–pectin
				$3.3 \geq 1.2$	Orange juice
				0.40	$0 \geq 10$ mM myristate
Small intestine	0.91	350–400 mL (0.5)	8 µL/min/kg (34 mL/hr)	(3–6 hr τ) 0.05–0.10	Olive oil (^{99}Tc) (@ 70 kg BW)
Cecum	0.02	–		(6–8 hr τ) 0.03–0.05	
Large intestine	0.51	1800 mL (2.5)		(10–60 hr τ)	Active → sedentary

Abbreviations: BW, body weight; GI, gastrointestinal.
[a] *Source:* Tissue weights and lumen volumes derived from Refs. 16, 19, 33, 40–43.
[b] *Source:* Transit times (τ) derived from Refs. 33, 36, 41–46.

mass balance from excreta, flows of urine, feces, and other pathways must be known to accurately account for all residues.

Urine flows (Q_u) follow a daily pattern; they are dependent on water consumption and a variety of other factors, which include level of physical activity, types of foods consumed, and exposure to drugs which may have direct effects on excretion of water by the kidneys. Human urine production decreases on a per kg BW basis during an ordinary lifetime, from 40 to 50 mL/kg BW day during the first two months of postnatal life, 30 to 40 mL/kg BW day in children, and drops to 10 to 25 mL/kg BW day in adults. This is approximately equal to the normal liquid consumption of 1500 mL/day for a 60-kg adult.

Feces production is, like urine production, dependent on the amounts of solid food taken in, content of digestible carbohydrates, protein, and fat. Absorption of water from the digestion process is relatively constant, except in disease. Overall, fecal production is closely related to the GI clearance parameters discussed above.

Inhalation and Exhalation vs. Absorption

The mass of a chemical available for exchange between the respiratory system and the air is assumed to equal the concentration of the chemical in the environmental air multiplied by the volume of air inhaled. The volume inhaled (tidal volume, V_t) is usually converted to a "minute volume," V_E, which is equal to the alveolar ventilation rate, Q_A, plus the dead space, DS:

$$Q_A = f(V_t - DS) \quad \text{or} \quad V_E = Q_A + f DS$$

The basal respiratory frequency f differs dramatically between species. This affects the average residence time for chemicals in all parts of the respiratory tract:

Species	f (bpm at rest)	Residence time $(\tau/2)$ (sec)	Reference
Human	8–15	2–3.7	(47)
Dog (beagle)	10–17	1.7–3	(48)
Rat	90–110	0.27–0.33	(49)
Mouse	200–250,	0.12–0.15,	(50,51)
	108–330	0.09–0.27	

The respiratory frequency also changes over a range characteristic for each species in response to physical activity and level of oxygenation of the blood.

Different experimental subjects will inhale differing amounts depending on species, body size, and level of cardiac exertion. Average values for alveolar ventilation (Q_A) established by Brown et al. (9) and Edwards et al. (52) for adult animals are as follows:

Alveolar Ventilation Rates at Rest

Species	Q_A (mL/min/kg)	Whole animal estimate
Mouse	1160	29 mL/min/25 g
Rat	529	132 mL/min/250 g
Rabbit	500	1000 mL/min/2 kg
Human	50	3.25 L/min/65 kg

The dose inhaled (per kg BW) during an experiment is obtained by multiplying concentration (C_i, mg/mL), exposure time (t), and Q_A. For a six-hour exposure, in the mouse these values would be

Dose inhaled $= Q_A C_i t = C_i$ (60 min/hr) (6 hr) (1160 mL/min/kg)
$= C_i$ (417,600 mL/kg).

These calculations do not yield the systemic dose. Total absorption (F) must be estimated from amounts excreted, by plasma (inhalation AUC)/(IV AUC) ratios or via partition coefficients using an inhalation model like that of Gerde and Dahl (53). The estimation of systemic absorption (F) by the inhalation route is complicated by the fact that significant exhalation of volatile chemicals is completed during each breath. As a consequence, sampling methods which cannot separate exhalation from inhalation underestimate absorption. Models which assume unidirectional flow, and do not correct for exhalation partitioning, generally overestimate the amount absorbed (53). Lumping of parameters and incorporation of multiple correction factors in lieu of measuring and incorporating important physiological processes in models may lead to misinterpretation of experimental results and poor model predictive power Anderson et al. (8).

The simplest inhalation exchange models assume rapid and uniform distribution of inhaled substances within the lung tissue and alveolar space. Differential equations describing absorption via this POE are similar to those for oral absorption, except that elimination may also occur, and passive diffusion (partitioning: $R_{blood:air} = C_{blood}/C_{air}$) is important:

$$\frac{dX_{lung}}{dt} = V_E C_{air} - Q_{lung} C_{blood}/R_{blood:air} \tag{12}$$

Here the flow Q_{lung} is the blood flow through the lungs, which is practically equivalent to the cardiac output (CO). Experimental measurement of exchange and systemic absorption of inhaled substances have shown that the assumption of rapid and uniform distribution holds only for very small molecules, such as H_2, CO_2, and oxygen. This is partly due to the short residence time of inhaled air, when compared to the diffusion velocity of molecules. The absorptive capacity provided by nasal tissue and the resistance presented by both nasal and alveolar tissue to systemic absorption is of more consequence than diffusion in air. Dahl et al. (48) have shown that nasal extraction is directly proportional to $R_{blood:air}$, but noted that alveolar extraction is always lower than predicted by Equation 12 and becomes saturated at about 40% of the inhaled dose. As a consequence, a permeability (P_{eff}) or an extraction efficiency (ε) parameter must be added to this equation in each direction if good fits to initial uptake curves are to be achieved.

REGULATORY APPLICATION OF PBPK MODELS

PBPK models have been constructed for a large number of chemicals over the past 20 years, for investigation of drug bioavailability (54,55), pharmacodynamic effect mechanisms (56), and a wide range of other purposes (57). All of these models ultimately have had regulatory implications, but the use of PBPK models by regulatory agencies as guidance for the resolution of questions raised in the review process, or as substitutes for additional experimental data, has been limited.

Construction of a useful pharmacokinetic or pharmacodynamic model requires the incorporation of equations that describe the POE into a systemic pharmacokinetic model, which may be simple or more complex, depending on the organs affected by the chemical of interest and the complexity of its metabolism. Pharmacokinetic models used for regulatory purposes must be tailored to the specific problem of interest, because real world problems are generally complex, and the amount of data and time available to solve them are limited. Figure 10 shows the pharmacokinetic compartments of a model depicting all major organs and POEs. Many of the organs shown in this model could be lumped into more generalized compartments depending on the questions asked and the data available, on which simulations and analyses will be based. The background and examples which follow illustrate models that have been developed for regulatory applications.

Oral Bioavailability

The factors that can influence oral absorption are physicochemical properties, physiological parameters, and the matrix in which the chemicals are delivered. Physiological factors which affect oral absorption were discussed in this chapter. Physicochemical properties include lipophilicity, solubility, diffusivity, chemical stability, pK_a, and particle form. Physiological characteristics which relate to the chemical itself are tissue permeability and active transport mechanisms. Chemical-specific parameters such as permeability are measured in vivo or in vitro [cell culture permeability (58,59)].

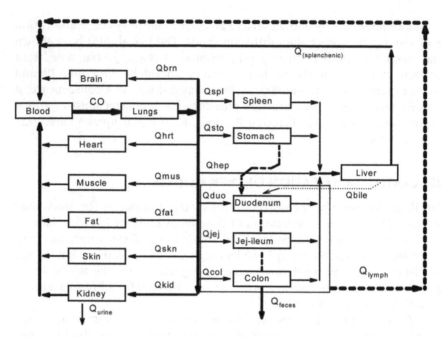

Figure 10 Schematic for a multicompartment, physiologically based model, including the gastrointestinal tract, with options for portal and lymphatic absorption. All fractional blood flows are indicated by symbols Q"i", while the total cardiac output is designated as CO. The parameter Qhrt represents only coronary artery flow. Lymphatic drainage is represented by the dashed line surrounding the intestines (19,27). Note that this model contains no compartments for tracking metabolites or metabolic reaction products (DNA or protein adducts, bound ligand, etc.). Organ volumes (V_j) for the rat are listed in Table 2. Gastrointestinal volumes and emptying rates are listed in Table 3. Blood flows and respiratory parameters may be found in Ref. 9. *Abbreviation*: CO, cardiac output.

Solubility and Drug Dissolution Models

Chemicals are not absorbed from the GI tract unless they are dissolved or emulsified to particle sizes smaller than 10 μm (60). Considerable effort is invested in the study and design of drug dissolution and absorption properties, because a slow dissolution rate or low solubility can prevent absorption of an otherwise effective drug. Several monographs review these topics in detail (61,62).

Lipophilic chemicals will generally dissolve or partition into a lipid/ hydrophobic phase from the aqueous phase of chyme in the intestinal lumen. If the lipids are digestible, absorption of lipid-soluble chemicals will accompany the uptake of micelles (bile acid–lipid complexes) as they are transported into the chylomicron/lymphatic pathway (63). Absorption of lipid-soluble chemicals therefore parallels that of digestible lipids, except in cases where the particle size of the phase is larger than that permitted by lymphatic uptake. When nondigestible phases (e.g., mineral oil, sucrose esters) are present in the intestines, partitioning of hydrophobic chemicals into these phases may prevent their absorption, and may cause the extraction of some chemicals from intestinal mucosal secretions that would otherwise be resorbed (64). A variety of vitamin and therapeutic drug delivery systems have been designed to exploit the characteristics of hydrophobic chemicals and the lymphatic uptake system (63).

Ionization state, pH, and strong binding to macroscopic components of a food matrix may cause the chemical of interest to be unavailable for absorption. Weak acids and weak bases may ionize in different portions of the GI tract. For materials absorbed by passive diffusion, the un-ionized chemical will be the species available for absorption. This fraction (F_{ui}) can be estimated from the Henderson–Hasselbach equation:

$$F_{ui} = 1 - F_i = 1 - \frac{1}{1 + 10^{(pK_a - pH)}} = \frac{10^{(pK_a - pH)}}{1 + 10^{(pK_a - pH)}} \tag{13}$$

When the absorption of a chemical can be changed by pH, model fit and parameter estimation results may be improved by converting total concentrations [$C(t)$] mentioned elsewhere in this section into "available" concentrations ($C_a = C F_{ui}$), using chemical-specific pK_a values and site-specific pH values to compute F_{ui}.

Dissolution is generally evaluated in vitro, using saline, simulated gastric, and/or intestinal fluids. The U.S. Pharmacopia has standardized formulas and procedures for determining dissolution. One of the simplest mathematical models of dissolution is the Noyes-Whitney (65) or Nernst–Brunner diffusion layer model. It assumes that there exists a thin layer of soluble chemical between the bulk of the dissolving fluid and the nonsoluble surface of undissolved particles for which transport into the bulk phase can be described:

$$\frac{dM_d(t)}{dt} = \frac{DA}{h}(C_s - C(t)) = K_s(C_s - C(t)) \tag{14}$$

where $M_d(t)$ is the mass of solute dissolved at time t [$M_d = C(t) V$], $C(t)$, the concentration in the bulk fluid [$C(t) = M_d/V$], C_s, the solubility of the solute, D, the diffusion coefficient in solution, A, the effective surface area of the undissolved solute, h, the thickness of the diffusion layer, and K_s is the dissolution rate coefficient.

If $M_i(0) \gg C_s V$, the surface area will remain approximately constant. Using the mass balance $M_i(0) = M_d(t) + M_i(t)$ to set boundary conditions, the following solutions are obtained for this idealized problem:

$$M_d(t) = C_s V [1 - e^{-(K_s/V)_t}] \quad M_i(0) > C_s V$$
$$M_i(t) = M_i(0) - C_s V [1 - e^{-(K_s/V)_t}] \tag{15}$$

In these solutions $M_i(t)$ is the amount of undissolved chemical at any time t, and V is the volume of solution with which the mass $M_i(0)$ was mixed. $M_i(0)$ is substituted for $C_s V$ when it is smaller of the two values. It should be noted from the equations above that the mass of chemical dissolved at equilibrium is dependent on the volume (V) of the solution, which is normally fixed during dissolution experiments in vitro, but increases with intestinal secretions in vivo.

A large number of factors which affect the dissolution rate are not incorporated into this formula, including temperature, pH, ionic composition, viscosity, and rate of stirring. If a standardized simulated intestinal fluid is used for dissolution rate measurement (i.e., USP methods) most of these variables could be ignored. The matrix or formulation in which the chemical is delivered as well as the geometry of dosage forms can have significant effects on its dissolution. Individual variation and diet can also affect many of these parameters in vivo.

Several authors (66,67) define A in terms of the number of particles per unit volume and the particle radius (r_p). On the basis of hydrodynamic arguments, Oh et al. (68) also use r_p as the limiting value for h, so that $K_s = (DA/r_p)$ in a more complex model of bioavailability described in the paper. For spherical particles it can be shown that the surface area is proportional to the power 2/3 of the particle weight. The mass of a sphere (m) can be defined in terms of its volume and radius:

$$m = \frac{4}{3} \pi r_p^3 \rho \qquad r_p^3 = \frac{3m}{4\pi\rho} \tag{16}$$

and its surface area $A = 4\pi r_p^2$. In considering a mass (M_i) containing N_t particles, the surface area for N_t particles can be estimated in terms of the mass, density (ρ), number of particles, and radius, or without explicit incorporation of the radius, which shrinks with dissolution:

$$A = 4\pi r_p^2 N_t = 4\pi N_t \left[\frac{3M_i}{4\pi\rho N_t} \right]^{2/3} \tag{17}$$

The dissolution rate can then be expressed in terms of solubility and the dissolved mass associated with the starting mass ($M_i = M_0 - M_d$)

$$\frac{dM_{d(t)}}{dt} = \frac{D}{h} \left[\frac{3M_i}{4\pi\rho N_t} \right]^{2/3} \left[C_s - \frac{M_d}{V} \right] \tag{18}$$

If the radius of the particle diffusion boundary is used to replace h (65,67) the expression for the radius returns with $M_d(t)$ to the power $1/3$

$$\frac{dM_{d(t)}}{dt} = 4\pi N_t D \left[\frac{3(M_0 - M_d(t))}{4\pi\rho N_t} \right] \left[C_s - \frac{M_d(t)}{V} \right] \tag{19}$$

The Hixson–Crowell "cube-root law" uses this assumption to express the dissolution rate K_s in terms of the particle weights after a (short) dissolution time t

$$K_s = \frac{M_0^{1/3} - M_d(t)^{1/3}}{t} \tag{20}$$

A variety of derivations for the dissolution rate incorporate the aqueous diffusion coefficient (D) which is rarely measured, and particle size distribution $\eta(r_t)$, which changes during dissolution (66,67). Many studies and models describe the effects of polymeric films (69) or diluent matrices (70). These derivations result in dissolution models with terms of $t^{1/2}$ (Fickian diffusion case) or t^n (non-Fickian dissolution) depending on starting assumptions and the particular formulation.

An extensive review of these theoretical models was published by Narasimha and Peppas (71). In most cases theoretical models are dispensed with in favor of empirical (fitted) dissolution functions, which are used as input to an oral absorption pharmacokinetic model. Linear, quadratic, and logistic models could be used as input functions (72), but will generally not incorporate adjustable parameters related to the dose $[M_i(0)]$ or volume (V). The Hixson–Crowell "cube-root law" [Eq. 2] provides an empirical parameter (K_s) which can be used in Equation (15) or more complex expressions which incorporate dose and volume to solve for the mass of chemical dissolved at any time t.

Much of the theoretical literature on dissolution is devoted to finding dissolution times and correlations between measurements made in vitro and estimates of bioavailability obtained from studies conducted in vivo. Useful correlations have only been found for a small range of solubility and permeability properties. A simple classification system has been developed to predict the usefulness of data from in vitro experiments in predicting bioavailability (72,73). The Biopharmaceutical Classification System (BCS) divides substances into four major categories on the basis of solubility and intestinal permeability properties (Table 4). Under this scheme a drug is considered to be highly soluble when the maximum dose is soluble in 250 mL (one cup) of water in the pH range of 1 to 7.5 (stomach $\rightarrow$ intestinal pH). It would be considered to be rapidly dissolving when at least 85% of the dose dissolves within 30 minutes (71).

Permeability is defined in terms of the fraction absorbed (F) in vivo, rather than in vitro (P_m).

Table 4 Biopharmaceutic Classification System

	High solubility (1 < pH < 7.5)	Low solubility
High permeability (F > 90%)	Absorption may be GI emptying rate independent	Likely to exhibit dissolution rate dependency
Low permeability (F < 90%)	Likely to exhibit GI emptying rate dependency	Likely to be poorly absorbed

Abbreviation: GI, gastrointestinal.

The BCS matrix has been further divided into four classes of expectations for in vitro–in vivo correlations (IV–IVC) which may be useful in planning (or skipping) such studies (Table 5).

It can be seen from these tables that in vitro dissolution results are likely to be predictive of in vivo bioavailability/absorption only for chemicals in Class II. Chemicals in Class I should have high bioavailability, but quantitative predictions will be poor if based on dissolution alone.

Example 1: Application of PBPK Models to Drug Screening and Clinical Study Design

Drug development is a very expensive process, which generally requires many years of testing after which only 10% to 20% of the lead candidate drugs are marketed. While many factors contribute to the high failure rate, pharmacokinetic properties, such as poor bioavailability, account for almost half of the failures. The difficulty of predicting bioavailability and efficacy in humans is due to the number of physiological and physicochemical variables involved in drug dissolution and absorption.

Yu (74) has developed a compartmental absorption and transit (CAT) model, a PBPK model that incorporates GI physiology (k_{sto} and k_{tr}) based on clinical administration situations and dissolution equations employing in

Table 5 IV–IVC Expectations

Class	Solubility	Permeability	IV–IVC expectation
I	High	High	Good if dissolution slower than gastric emptying, otherwise none
II	Low	High	Dissolution rate limited
III	High	Low	Absorption is rate limiting and no IV–IVC is expected
IV	Low	Low	Limited or no IV–IVC expected

Abbreviation: IV–IVC, in vitro–in vivo correlation.

vitro dissolution and permeability measurements to predict the absorption of new drugs in clinical studies. It has also been used as an in silico method for screening new drugs and designing clinical studies. Examples of drugs used to evaluate this model include digoxin (a cardioactive drug), griseofulvin (an antifungal agent), and cefatrizine (a β-lactam antibiotic). Each of these drugs was known to have bioavailability problems, but the limiting factors were not known and a model was needed which could evaluate the complex interactions of dissolution, intestinal transport, and permeability in the absorption process.

The model consists of nine volumes, of which seven represent the lumen of the small intestine. Each of the small intestinal volumes is divided into two compartments, one of which contains an undissolved drug (M_i), and the other dissolved drug (M_d). Additional subscripts are associated with each volume in the representative equations (M_{ni} or M_{nd}) to indicate which of the seven successive small intestinal volumes and states of dissolution is being represented (Fig. 11).

The term ($2P_{eff}/R$) could be replaced by $K_{a(n)}$ (absorption rate coefficient)—the choice is author dependent, and in this case is the result of conversion factors used to convert from permeability coefficients measured in vitro to parameters useable for the in vivo model. The solid drug particles here are assumed to be smaller than 1 mm—the size limit above which

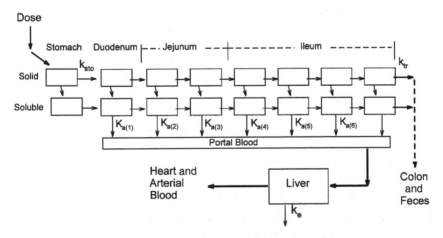

Figure 11 CAT model of Yu (74). The solid dosage form is delivered to the stomach, which is assumed to have an absorption coefficient (K_a) equal to zero. In intestinal compartments the absorption coefficients are computed from segment-specific permeabilites: $K_{a(n)} = 2P_{eff\ (n)}/R_n$, where R^n is the radius of the segment. The solid mass in each segment (M_{ni}) can dissolve to enter the soluble compartment of the segment (M_{nd}), which is available for absorption. Systemic tissues can be added but only a single (central) compartment was part of the original model (equations in Table 6). *Abbreviation*: CAT, compartmental absorption and transit.

stomach emptying becomes nonlinear. Stomach emptying of solutions and small suspended particles can be accurately modeled as a first-order process (K_{sto}), as is the emptying of the small intestine (k_{tr}). In the case of cefatrizine, absorption is known to be mediated primarily via active transport. Cefatrizine absorption was therefore modeled with the inclusion of a saturable permeability function, as shown in Table 6. A numerical integration algorithm (ADAPT) was used to solve the differential equations using input parameter values, as shown in Table 7.

The diffusion layer thickness (h) was assumed to be 30 μm and density (ρ), 1.2 g/cm^2, while particle radii were varied from 5 to 100 μm to determine drug characteristics limiting the dissolution. The mass of drug in the intestinal mucosa and serosal tissue is ignored in this model. Transport is assumed to occur directly from the intestinal lumen into the portal blood.

Results

Several additional models have been developed (67,75), and a more complex model derived from the model of Yu (74) is currently available as a commercial software product (GastroPlusJ, Simulations Plus, Inc.).

When compared to simple dissolution models or one-compartment (gut) absorption models, the PBPK model of Yu (74) is quite realistic and allows simulation of complex GI transit and absorption processes that would not be resolvable by "thought experiments" and would be very difficult to analyze by animal experiments. It also allows one to have greater use of in vitro study results (vs. extensive animal studies) and more confidence in clinical studies planned through the use of this tool. Specific results were as follows:

Digoxin—The model predicted that drug particles of size 8 μm or less would be completely dissolved and absorbed, while larger particle sizes

Table 6 Differential Equations for Dissolution/Permeation Model

$$\frac{dM_{sto,i}}{dt} = U(t) - k_{sto}M_{sto,i} \qquad \text{STO = First compartment } (n = 1)$$

$$\frac{dM_{sto,d}}{dt} = U(t) - k_{sto}M_{sto,d} \qquad k_{sto} = 2.0/\text{hr} \qquad k_{tr} = 0.20/\text{hr}$$

All intestinal volumes ($n = 2$–7) have the following generic equations:

$$\frac{dM_{ni}}{dt} = k_{tr}(M_{(n-1)i} - M_{ni}) = \frac{3DM_{ni}}{\rho hr}\left(C_g - \frac{M_{nd}}{V_n}\right)$$

$$\frac{dM_{nd}}{dt} = k_{tr}(M_{(n-1)d} - M_{nd}) + \frac{3DM_{ni}}{\rho hr}\left(C_s - \frac{M_{nd}}{V_n}\right) - \frac{2P_{eff}M_{nd}}{R}$$

$$\frac{dM_{systemic}}{dt} = \left(\frac{2P_{eff}}{R}\right)\sum M_{nd} - k_e M_{systemic}$$

$$K_{a,n} = \frac{V_{max}}{(K_{m,n} + [M_{nd}/V_n])}\left(\text{replaces } \frac{2P_{eff}}{R}\right)\text{(saturable uptake for cefatrizine)}$$

Table 7 Parameters for PBPK Dissolution/Permeation Model

Parameters	Digoxin	Griseofulvin	Panadiplon
Dose [$U(t)$] (mg)	0.5	500	10
Solubility (C_S) (μg/mL)	24	15	77
Diffusion coefficient (D) (cm²/sec)	1.5×10^{-6}	8.2×10^{-6}	6.1×10^{-6}
Permeability (P_{eff}) (cm/sec)	1×10^{-3}	1.6×10^{-3}	0.185×10^{-3}

Abbreviation: PBPK, physiologically based pharmacokinetic.

would fail to dissolve completely during the transit time. These predictions were in agreement with the results from three independent studies.

Griseofulvin—The model predicted that absorption of this drug would be limited both by dissolution time and solubility. The maximum absorption fraction (F) was predicted to be 38% at a dose of 500 mg, which is close to the experimental value of 45%.

Cefatrizine—As shown in Figure 12, the CAT model, including a Michaelis–Menten submodel of saturable permeability, was very successful

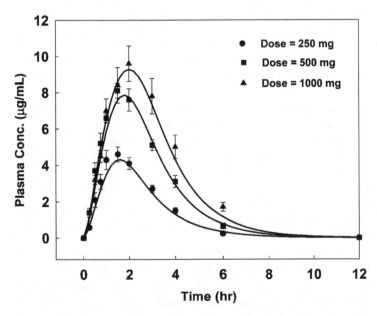

Figure 12 Effect of increasing dose on absorption of cefatrizine. The fraction absorbed (F) decreases with increasing dose because of saturation of the (*active*) transporter. CAT model simulations are represented by solid lines, while measured plasma levels are shown by solid symbols. *Abbreviation*: CAT, compartmental absorption and transit. *Source*: From Ref. 54.

in predicting the declining fraction of cefatrizine absorbed with increasing absorption. Previous models, including saturable absorption but without intestinal transport, predicted much greater absorption than the CAT model.

Example 2: PBPK Model Prediction of Intestinal Bacterial Metabolism

Cyclamate (CM) is an artificial sweetener which was banned by the FDA in 1969 after bladder tumors were found in rats during chronic studies (76). After CM was banned, other toxic effects were noticed, such as reproductive (77) and potential cardiovascular effects, but were not the basis for regulatory action. Tumors caused by CM have been shown to be the result of crystallization of Na^+ CM in the bladder, a phenomenon which has been observed for naturally occurring acids, such as ascorbic acid (vitamin C) when fed in large amounts as the sodium salt. This new information has resulted in efforts to reinstate the use of CM as an artificial sweetener.

CM is known to be metabolized to varying extents by intestinal bacteria [clostridia in rats, enterococci in humans (78)]. The product of this metabolism, cyclohexylamine (CHA), is absorbed rapidly, and has been shown to cause cardiovascular effects [increased blood pressure (BP)] in clinical studies (79). Occasional exposure to CM does not result in the formation of large amounts of CHA (80). However, under conditions of continuous exposure, bacterial conversion capacity is induced and may reach levels higher than 60% in a few individuals (81).

The clinical study data available for CM and CHA were initially too limited to allow direct determination of the likelihood of increasing BP in a significant number of CM consumers. A pharmacokinetic modeling effort was conducted to (*i*) fit the available data for absorption and elimination of both CM and CHA and (*ii*) predict plasma levels of CHA given that the bacterial conversion of CM to CHA from 0% to 50%. Predicted plasma levels can be used with BP data from the clinical study with CHA to predict changes in BP as a function of CM intake (Figs. 13 and 14).

Physiological parameters for the GI tract (k_{sto} and k_{tr}) were initially set to values typical of humans consuming a normal diet (solid food). Intestinal transit (k_{tr}) was then adjusted to move CHA production into time periods consistent with clinical results and normal cecum transit. Saturable metabolism was incorporated into the pharmacokinetic model with modification of the CHA production equations by replacing first-order k_{32} with k_{met}, as shown in Table 8. Simulations were performed under several dosing scenarios to determine the effect of saturable metabolism on the rate of CM $\rightarrow$ CHA conversion (k_{met}) and predicted plasma CHA levels.

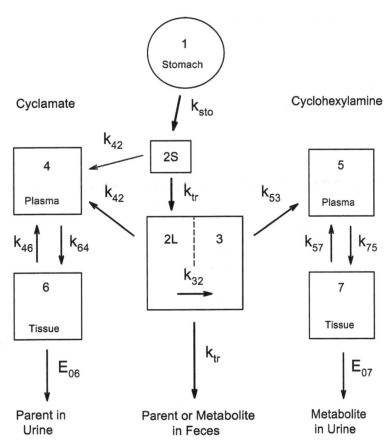

Figure 13 Schematic of model used to simulate intestinal bacterial production of CHA from CM. Plasma–tissue transport parameters were adjusted to known CM absorption (40%) half-life and CHA tissue:plasma partitioning (10:1). As described in the text and shown in Figures 14 and 15, the intestinal conversion and transport coefficients (k_{32} and k_{tr}) were adjusted to achieve plasma levels of CHA observed in clinical studies by Eichelbaum et al. (79) and Buss et al. (81). *Abbreviations*: CHA, cyclohexylamine; CM, cyclamate.

Results

Under the dosing conditions employed in the clinical studies mentioned above, CHA peaks of 1- to 2-µg/mL plasma could be attained in subjects with episodic conversion capacities of 50% or more. The saturability of the bacterial converting enzyme made only a small difference (20% less) in the fraction of CM converted to CHA at high CM doses (> 0.5 g/subject), contrary to the expectations of the study sponsors (Fig. 15). Although the predicted CHA levels of 1- to 2-µg/mL plasma could change BP by as much as 10 mmHg, the intrinsic

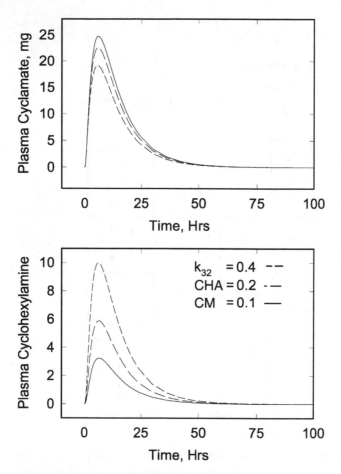

Figure 14 Simulations conducted to estimate the intestinal bacterial conversion coefficient (CM → CHA). The conversion rate coefficient (k_{32}) was estimated by adjusting k_{32} with other parameters fixed until CHA conversion reached the desired percent. Note the reduction of plasma CM levels imposed by the overall mass balance. *Abbreviations*: CHA, cyclohexylamine; CM, cyclamate.

variability in resting BP for individuals documented in independent studies was significantly greater than 10 mmHg (82).

Inhalation and Dermal Absorption Models

General Form of Models

PBPK models in use for simulation of inhalation and dermal routes of exposure generally assume that blood–tissue distribution of the chemical of interest is perfusion limited. These models may contain relatively few volumes (compartments) for distribution if metabolism and elimination

Table 8 Differential Equations Describing CM to CHA Conversion

$\frac{dM_1}{dt} = \text{Input}(t) - (k_{4,1} + k_{\text{sto}})M_1$ $\quad k_{\text{sto}} = 0.5/\text{hr} \quad k_{\text{tr}} = 0.25/\text{hr}$

$\frac{dM_{2S}}{dt} = k_{\text{sto}}M_1 - (k_{4,2} + k_{\text{tr}(S)})M_{2S}$ $\quad$ (CM from small intestine)

$\frac{dM_{2L}}{dt} = k_{\text{tr}(S)}M_{2S} - (k_{3,2} + K_{4,2} + k_{\text{tr}(L)})M_{2L}$ $\quad$ (CM absorption and conversion)

$\frac{dM_3}{dt} = k_{3,2}M_3 - (k_{5,3} + k_{\text{tr}(L)})M_3$ $\quad$ (CM $\rightarrow$ CHA, $k_{3,2}$ first order)

$\frac{dM_4}{dt} = k_{4,2}(M_{2S} + M_{2L}) - k_{6,4}M_4 + k_{4,6}M_6$ $\quad$ (plasma CM)

$\frac{dM_5}{dt} = k_{5,3}M_3 + k_{5,7}M_7 - k_{7,5}M_5$ $\quad$ (plasma CHA)

$\frac{dM_6}{dt} = k_{6,4}M_4 - (k_{4,6} + E_{06})M_6$ $\quad$ (tissue CM)

$\frac{dM_7}{dt} = k_{7,5}M_5 - (k_{5,7} + E_{07})M_7$ $\quad$ (tissue CHA)

Saturable metabolism submodel (k_{met} replaces $k_{3,2}$):

$k_{\text{met}} = \frac{k_p}{(K_m + [S])}$

$\frac{dM_{2L}}{dt} = k_{\text{tr}(S)}M_{2S} - (k_{\text{met}} + k_{4,2} + k_{\text{tr}(L)})M_{2L}$

$\frac{dM_3}{dt} = k_{\text{met}}M_{2L} - (k_{\text{tr}(L)} + k_{5,3})M_3$

where

[S] intestinal lumen (cecum) concentration of CM
K_m Michealis constant for CM $\rightarrow$ CHA converting enzyme
k_p maximum conversion rate/mL chyme (V_{max}/unit volume)
$k_{\text{tr}(i)}$ are transport rates describing the transfer of chyme from one intestinal compartment to the next

Abbreviations: CM, cyclamate; CHA, cyclohexylamine.

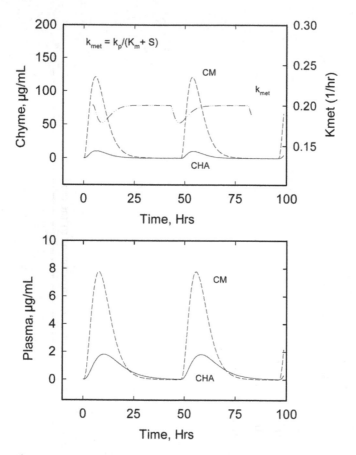

Figure 15 Effect of intestinal CM concentration on k_{met} and CHA production. At a dose of 1 g CM/subject, CM is diluted by stomach and intestinal fluids to the point that decrease of k_{met} by substrate saturation of converting enzyme is less than 25%. *Abbreviations*: CHA, cyclohexylamine; CM, cyclamate.

are simple. GI physiology can be largely ignored, except in cases where enterohepatic recycling is important (83,84), or where route-to-route extrapolation (oral to inhalation) is the primary use of the model as described above.

Inhalation exposure models generally assume that there is little or no barrier to the entry of the inhaled chemical into lung tissue and systemic distribution. Development of inhalation exposure models generally emphasizes the measurement of blood:air and tissue:blood partition coefficients, because lung tissue must be kept hydrated and viable to measure a lung:air partition coefficient. The difficulty of maintaining the lung:air preparation has resulted in the use of a ratio of measured partition coefficients as a surrogate for the lung:air coefficient:

$$\langle R_{\text{lung:air}} \rangle = \frac{R_{\text{lung:blood}}}{R_{\text{air:blood}}} \tag{21}$$

Protocols for measuring these parameters have been developed by several laboratories, and tend to be chemical specific. The first section describes complications which may be important in estimating partition coefficients. These limitations are often the major determinants of the accuracy of PBPK models for volatile chemicals (85,86). Simple anatomical (tissue volumes) and physiological parameters (blood flows and minute volumes) incorporated into inhalation exposure models are generally standardized for test animals and humans (9). Considerable effort has been invested in the development of more detailed, species-specific anatomical parameters for inhalation models particularly those dealing with particulates (48). Methodology for standardization and application of inhalation dosimetry for regulatory purposes has been published by the U.S. Environmental Protection Agency (EPA) (3).

Dermal exposure models are based more on the concept of permeability, because the stratum corneum presents a very significant, and often effectively impermeable barrier to the entry of chemicals into lower levels of dermal tissue and systemic distribution. Methods for measurement of percutaneous absorption have been standardized (87,88), and are the primary parameters used to control absorption in PBPK models for this POE. Dermal permeability is described by Fickian diffusion equations:

$$\frac{dX}{dt} = J(t)A = \frac{DA}{h}(C_a - C_b) = P_{\text{eff}}(C_a - C_b) \tag{22}$$

where X is the mass of chemical passing through the barrier from side "a" to side "b," $J(t)$, the flux (mass of chemical) passing through a unit area of barrier, D, the diffusion coefficient (cm^2/sec) for the chemical through the barrier material, h, the thickness (cm) of the barrier, A, the total area (cm^2) of the barrier, C_a and C_b are the concentrations of the chemical on respective sides "a" and "b," and P_{eff} is the effective permeability.

Dermal exposure models have been used extensively to estimate occupational absorption of industrial solvents, cleaning materials, and liquid fuels. They have also been used to estimate systemic exposure to cosmetics and dermally applied drugs.

Example 3: Application of PBPK Models to Hazardous Air Pollutant Rulemaking

The EPA Office of Pollution Prevention and Toxics has recently used PBPK modeling as part of a series of Enforceable Consent Agreements (ECA) in an effort to improve the quality of data used in hazardous air pollutant rulemaking. The use of PBPK models allows the Agency to attain several goals simultaneously:

1. PBPK modeling provides guidance to study sponsors concerning the most critical parameters, time points, and experimental error sensitivity before they embark on new studies (use existing information to improve efficiency of use of study resources).
2. PBPK models allow the agency to perform detailed, scientifically sound transformations of data from oral dosing studies to equivalent inhalation exposures (to develop computational modeling approaches for quantitative route-to-route extrapolation).
3. PBPK models provide a rational, mathematical means of integrating mechanistic data obtained from in vitro and limited in vivo studies into whole animal models from which results of larger, more expensive traditional studies (subchronic and chronic) can be predicted.
4. PBPK modeling allows the Agency to implement guidance from Interagency Coordinating Committee on the Validation of Alternative Methods on the use of laboratory animals in testing (reduce, replace, and refine) through (*i*) broader use of existing studies, (*ii*) development and refinement of mechanistic data, and (*iii*) development and use of computational modeling approaches for quantitative route-to-route extrapolation.

The particular case described here is for the industrial chemical 1,2-dichloroethane (DCE). A considerable body of experimental data on the toxicity, metabolism, and pharmacokinetics has been available for many years (89,90). PBPK models had been developed, and suggested a minimal set experiment needed to refine risk assessments by adding mechanistic information to extrapolate systemic doses observed to biochemical effects measured in independent experiments. Route-to-route extrapolation was used to estimate systemic doses in oral dosing studies on the basis of parameters from inhalation experiments.

The model used for the ECA on DCE is based on the model of D'Souza et al. (91), with modification of the GI tract parameters to include periodic consumption of water as an input source for DCE. The model predicts tissue concentrations of DCE and its metabolites, as well as the effects of this chemical on concentrations of glutathione (GSH) in the lung and the liver, where GSH levels are directly correlated with toxicity.

General DCE Model Structure

The model developed by D'Souza et al. (91,92) is parameterized in the same way as it was done in the earlier model to simulate styrene disposition by Ramsey et al. (95). Tissue volumes are lumped into three nonmetabolizing groups (slowly perfused, richly perfused, and adipose) plus two individual organs which can metabolize DCE. The lung serves both as a POE and as an eliminating organ. The liver serves only as a metabolizing organ. The GI tract (gut) is represented by a single absorption rate coefficient (K_a), otherwise not parameterized. The liver is subdivided into mathematical compartments which correspond to metabolites of DCE. The liver submodel also contains a function which simulates GSH synthesis and reaction losses. A schematic of the DCE model is shown in Figure 16.

Depletion of GSH is thought to be the mechanism for acute toxicity of DCE, while the mutagenic/carcinogenic properties are believed to be associated with the GSH adduct of DCE: 2-chloroethylglutathione (CE-SG) which may form DNA-reactive S-episulfonium ethylglutathione Reitz et al. (90). To keep a mass balance for DCE and its GSH-reactive metabolites, formation of the reactive metabolite chloroacetaldehyde (CAA) and its non-GSH containing metabolites must also be tracked by this model. Similar equations and submodel structures were added to the lung submodel to compare the significance of DCE metabolism by the lung to that predicted in the liver. Table 9 lists the differential equations incorporated into the model. Physiological (blood and respiratory flows) and anatomical (tissue volumes) parameters were constants in the simulations. Chemical-specific parameters were varied to achieve a good fit to observations.

Data Sources and Predictions

Chemical-specific parameters for EDC were based on parameters developed by Gargas et al. (93) for methylene chloride and Andersen et al. (94) and Reitz (90) for EDC. Model calibration data were obtained from Spreafico et al. (89). Model validation data are to be obtained through studies designed as part of the ECA for EDC.

Results

Simulations were conducted of dosing with DCE by the oral route for comparison with experimental measurements of GSH described in D'Souza et al. (91). Model predicted reductions of GSH after dosing (about 60% maximum reduction), and restoration of GSH levels on elimination of DCE was within 20% of the observed measurements of liver and lung GSH concentrations at all time points, for both rat and mouse simulations. Inhalation exposure resulted in lower systemic concentrations for an equivalent dose than did oral doses delivered in corn oil. Simulations of these two routes of exposure were consistent with the experimental observations and indicated that oral dosing resulted in a reduced loss of DCE by exhalation (in comparison with inhalation exposure) as well as significantly greater metabolism to potentially reactive metabolites. This finding (both experimental and by simulation) is

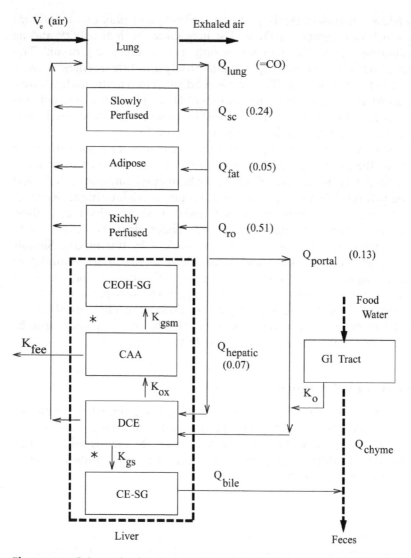

Figure 16 Schematic description of model of D'Souza et al. (91). The liver is expanded into five compartments (CEOH-SG, CAA, DCE; CE-SG, and GSH). The coefficient K_{fee} is associated with protein-imine formation and other cell component condensations. The compartment corresponding to GSH is not shown, but would provide input to reactions indicated with an asterisk (ρ). Exhalation and hepatic metabolism are the primary routes of elimination. *Abbreviations*: CAA, chloroacetaldehyde; CE-SG, 2-chloroethylglutathione; CEOH-SG, 2-chloro-1-hydroxyethylglutathione; DCE, 1,2-dichloroethane; GSH, glutathione.

Table 9 DCE Model Differential Equations

$$\frac{dX_{\text{lung}}}{dt} = Q_{\text{air}}(C_{\text{air}} - C_{\text{lung}}/R_{\text{lung:air}}) + Q_{\text{CO}}(C_{\text{various}} - C_{\text{art}}) - (K_{\text{ox}} + K_{\text{gs}}[\text{GSH}])C_{\text{lung}}V_{\text{lung}}$$

$$\frac{dX_{\text{sl}}}{dt} = Q_{\text{sl}}(C_{\text{art}} - C_{\text{sl}}/R_{\text{sl}}) \qquad \text{(slowly perfused)}$$

$$\frac{dX_{\text{fat}}}{dt} = Q_{\text{fat}}(C_{\text{art}} - C_{\text{fat}}/R_{\text{fat}})$$

$$\frac{dX_{\text{rp}}}{dt} = Q_{\text{rp}}(C_{\text{art}} - C_{\text{rp}}/R_{\text{rp}}) \qquad \text{(rapidly perfused)}$$

$$\frac{dX_{\text{sto}}}{dt} = \text{Input}(t) - K_{\text{a}}X_{\text{sto}}$$

$$\frac{dX_{\text{liver}}}{dt} = Q_{\text{hep}}(C_{\text{art}} - C_{\text{liv}}/R_{\text{liv}}) + K_{\text{a}}X_{\text{sto}} - \left(K_{\text{ox}} + \frac{K_{\text{gs}}[\text{GSH}]}{K_{\text{m}} + C_{\text{liv}}}\right)C_{\text{liv}}V_{\text{liv}}$$

$$\frac{dX_{\text{venous}}}{dt} = \sum_j Q_j(C_j/R_j) - Q_{\text{CO}}C_{\text{art}}$$

$$C_{\text{art}} = \frac{C_{\text{air}}Q_{\text{air}} + C_{\text{venous}}Q_{\text{CO}}}{Q_{\text{CO}} + Q_{\text{air}}/R_{\text{blood:air}}} \qquad \text{(assuming rapid equilibrium at steady state)}$$

Metabolite formation equations:

$$\frac{dY}{dt} = \left(\frac{K_{\text{gs}}[\text{GSH}]}{K_{\text{m}} + C_{\text{liv}}}\right)C_{\text{liv}}V_{\text{liv}}$$

$$\frac{d[\text{CAA}]}{dt} = K_{\text{ox}}C_{\text{liv}} - K_{\text{fee}}[\text{CAA}]$$

$$\frac{dZ}{dt} = \left(\frac{K_{\text{gsm}}[\text{GSH}]}{K_{\text{m}} + [\text{CAA}]}\right)[\text{CAA}]V_{\text{liv}}$$

$$V_{\text{liv}}\left(\frac{d[\text{GSH}]}{dt}\right) = K_0 + K_{\text{ind}}[\text{GSH}] - \left(\frac{dY}{dt} + \frac{dZ}{dt}\right)$$

Note: Masses denoted with an "X_j" represent the parent chemical—DCE. Other metabolites are denoted by "Y" = CE-SG, "Z" = CEOH-SG, and CAA.

Abbreviations: GSH, glutathione; DCE, 1,2-dichloroethane.

consistent with bioassay experiments which resulted in tumors via oral dosing (corn oil) but no tumors via inhalation. Future results will be used to validate the model for prediction of carcinogenicity at low doses.

REFERENCES

1. Roth WL, Young JF. Use of pharmacokinetic data under the FDA's Redbook II guidelines for direct food additives. Int J Toxicol 1998; 17:355–381.
2. Rozman KK, Klassen CD. Absorption, distribution and excretion of toxicants. In: The Basic Science of Poisons. 6th ed. McGraw-Hill, New York, NY: Casarett & Doull's TOXICOLOGY, 2001:107–132.
3. Jarabeck AM, Hanna L, Menache M, Overton J. Methods for Derivation of Inhalation Refererence Concentrations and Application of Inhalation Dosimetry. U.S. Environmental Protection Agency EPA/600/8-90/066F, 1994.
4. Gibaldi M, Perrier D. Pharmacokinetics. 2nd ed. New York: Marcel Dekker, 1982.
5. Benet LZ. General treatment of linear mammillary models with elimination from any compartment as used in pharmacokinetics. J Pharm Sci 1972; 61:536–541.
6. Wagner JG. Pharmacokinetics for the Pharmaceutical Scientist. Lancaster, PA: Tecnomic Publishers, 1993.
7. Notari RE. Biopharmaceutics and Clinical Pharmacokinetics. New York: Marcel Dekker, 1987.
8. Anderson ME, Clewell HJ, Frederick CB. Applying simulation modeling to problems in toxicology and risk assessment–a short perspective. Toxicol Appl Pharmacol 1985; 133:181–187.
9. Brown RP, Delp MD, Lindstedt SL, Rhomberg LR, Beliles RP. Physiological parameter values for physiologically-based pharmacokinetic models. Toxicol Indust Health 1997; 13(4):407–484.
10. Kidwell JF, Howard A, Laird AK. The inheritance of growth and form in the mouse. II. The gompertz growth equation. Growth 1969; 33:339–352.
11. Molnar JA, Alpert NM, Burke JF, Young VR. Relative and absolute changes in soluble and insoluble collagen pool size in skin during normal growth and with dietary protein restriction in rats. Growth 1987; 51:132–145.
12. Kurnick NB, Keren RL. The effect of aging on the desoxyribonuclease system, body and organ weight and cellular content. J Gerontol 1962; 17:245–253.
13. Charles River Laboratories. Price List and Product Guide. Massachusetts: Wilmington, 1996.
14. Cameron TP, Hickman RL, Komreich MR, Tarone RE. History, survival and growth patterns of B6C3F1 mice and F344 rats in the National Cancer Institute Carcinogenesis Testing Program. Fund Appl Toxicol 1985; 5:526–538.
15. Li X, Weber LWD, Rozman KK. Toxicokinetics of 2,3,7,8-tetrachlorodibenzo-dioxin in females sprague-dawley rats including placental and lactational transfer to fetuses and neonates. Fund Appl Toxicol 1995; 27:70–76.
16. Deavers S, Huggins RA, Smith EL. Absolute and relative organ weight of the growing beagle. Growth 1972; 36:195–208.

17. Friedman L, Gaines DW, Newell RF, et al. Growth patterns in selected organs of the miniature swine as determined by gross macromolecular composition. J Anim Sci 1995; 73: 1340–1350.
18. Donaldson HH. The Rat: Data and Reference Tables. Philadelphia, Pennsylvania: Wistar Institute of Anatomy and Biology, 1924.
19. Roth WL, Freeman RA, Wilson AGE. A physiologically based model for gastrointestinal absorption and excretion of chemicals carried by lipids. Risk Anal 1993; 13:531–543.
20. Pitts GC, Bull LS. Exercise, obesity and growth in the rat. Am J Physiol 1977; 232:R38–R44.
21. Yokogawa K, Nakashima E, Ichimura F. Effect of tissue volume on the distribution kinetics of biperiden as a function of age in rats. Drug Metab Disp 1990; 18:258–263.
22. Rescigno A, Segre G. Theory of compartment system models. In: Drug and Tracer Kinetics. Waltham, Massachusetts: Blaisdell Publishing, 1966:75–137.
23. Roth WL, Weber LWD, Rozman KK. Incorporation of first-order uptake rate constants from simple mammillary models into blood-flow limited physiological pharmacokinetic models via extraction efficiencies. Pharm Res 1995; 12(2): 263–269.
24. Lin JH, Sugiyama Y, Awazu S, Hanano M. In vitro and in vivo evaluation of the tissue-to-blood partition coefficient for physiological pharmacokinetic models. J Pharmacokin Biopharm 1982; 10(6):637–647.
25. Chen HG, Gross JF. Estimation of tissue-to-plasma partition coefficients used in physiological pharmacokinetic models. J Pharmacokin Biopharm 1979; 7(1):117–125.
26. Gallo JM, Lam FC, Perrier DG. Area method for the estimation of partition coefficients for physiological pharmacokinetic models. J Pharmacokin Biopharm 1987; 15(3):271–280.
27. Roth WL, Ernst S, Weber LWD, Kerecsen L, Rozman K. A pharmacodynamically responsive model of 2,3,7,8-tetrachlorodibenzo-p-dioxin (TCDD) transfer between liver and fat at low and high doses. Toxicol Appl Pharmacol 1994; 127:151–162.
28. Buelke-Sam J, Holson JF, Nelson CJ. Blood flow during pregnancy in the rat: II. Dynamics of and litter variability in uterine flow. Teratology 1982; 26:279–288.
29. Luecke RH, Wosilait WD, Young JF. Mathematical modeling of human embryonic and fetal growth rates. Growth, Development and Aging 1999; 63:49–59.
30. O'Flaherty EJ, Clarke DO. Pharmacokinetic/pharmacodynamic approaches for developmental toxicity. In: Kimmel CA, Buelke-Sam J, eds. Developmental Toxicology. 2d ed. New York: Raven Press, Ltd., 1994.
31. Roth WL, Garthoff LH, Luecke RH, Young JF. Standardized growth models of experimental animals for reconstruction of whole animal pharmacokinetics. Proceedings 10th International Conference on Mechanics in Medicine and Biology 1998:363–366.

32. Iatropoulos MJ. Morphology of the gastrointestinal tract. In Rozman K, Hänninen O, eds. Gastrointestinal Toxicology, Chapter 9. Elsevier Science, 1986:246–266.

33. Clemens ET, Stevens CE. A comparison of gastrointestinal transit time in ten species of mammal. J Agric Sci (Camb) 1980; 94:735–737.

34. Stevens EC, Hume ID. Comparative Physiology of the Vertebrate Digestive System. 2nd ed. The mammalian gastrointestinal tract, Chapter 4. Cambridge University Press, 1995.

35. Hunt JN, Knox MT. Regulation of gastric emptying. In: Heidel W, ed. Handbook of Physiology, Section 6: Alimentary Canal. Washington, DC: American Physiological Society, 1968:1917–1935.

36. Hinder RA, Kelly KA. Canine gastric emptying of solids and liquids. Am J Physiol 1977; 233:E335–E339.

37. Nimmo WS, Heading RC, Wilson J, Tothill P, Prescott LF. Inhibition of gastric emptying and drug absorption by narcotic analgesics. Br J Clin Pharm 1975; 2:509–513.

38. Wilding IR, Davis SS, Hardy JG, Robertson CS, John VA, Powell ML, Leal M, Lloyd P, Walker SM. Relationship between systemic drug absorption and gastrointestinal transit after the simultaneous oral administration of carbamazepine as a controlled-release system and as a suspension of ^{15}N-labeled drug to healthy volunteers. Br J Clin Pharmacol 1991; 32:573–579.

39. Dressman JB, Amidon GL, Reppas C, Shah VP. Dissolution testing as a prognostic tool for oral drug absorption: immediate release dosage forms. Pharm Res 1998; 15(1):11–22.

40. Hebel R, Stromberg MW. Anatomy of the Laboratory Rat. Baltimore, MD: Williams & Wilkins, 1976.

41. Kellow JE Gastrointestinal motility and defecation. In Greger R, Windhorst U, eds. Comprehensive Human Physiology, Chapter 63, Vol. 2. Springer-Verlag, 1996.

42. Brown NJ, Rumsey RDE, Read NW. Adaptation of hydrogen analysis to measure stomach to caecum transit time in the rat. Gut 1987; 28:849–854.

43. Thompson RC, Hollis OL. Irradiation of the gastrointestinal tract of the rat by ingested ruthenium-106. Am J Physiol 1958; 194(2):308–312.

44. Enck P, Merlin V, Erckenbrecht JF, Wienbeck M. Stress effects on gastrointestinal transit in the rat. Gut 1989; 30:455–459.

45. Read NW, Miles CA, Fisher D, Holgate AM, Kime ND, Mitchell MA, Reeve AM, Roche TB, Walker M. Transit of a meal through the stomach, small intestine, and colon in normal subjects and its role in the pathogenesis of diarrhea. Gastroenterology 1980; 79:1276–1282.

46. Williams CL, Villar RG, Peterson JM, Burks TF. Stress-induced changes in intestinal transit in the rat: a model for irritable bowel syndrome. Gastroenterology 1988; 94:611–621.

47. Frostell C, Pande JN, Hedenstierna. Effects of high-frequency breathing on pulmonary ventilation gas exchange. J Appl Physiol 1983; 55(6):1854–1861.

48. Dahl AR, Schlesinger RB, D'A. Heck H, Medinsky MA, Lucier GW. Comparative dosimetry of inhaled materials: differences among animal species and extrapolation to man. Fund Appl Toxicol 1991; 16:1–13.

49. Olsen EB, Dempsy JA. Rat as a model for humanlike ventilatory adaptation to chronic hypoxia. J Appl Physiol 1978; 44(5):763–769.
50. Vijayaraghavan R, Schaper M, Thompson R, Stock MF, Alarie Y. Characteristic modifications of the breathing pattern of mice to evaluate the effects of airborne chemicals on the respiratory tract. Arch Toxicol 1993; 67:478–490.
51. Fairchild GA. Measurement of respiratory volume for virus retention studies in mice. Appl Microbiol 1972; 24(5):812–818.
52. Edwards AWT, Korner PI, Thronburn GD. The cardiac output of the anesthetized rabbit, and the effects of preliminary anaesthesia, environmental temperature and carotid occlusion. Quart J Exptl Physiol 1959; 44(3):309–321.
53. Gerde P, Dahl AR. A model for the uptake of inhaled vapors in the nose of the dog during cyclic breathing. Toxicol Appl Pharmacol 1991; 109:276–288.
54. Yu LX, Ellison CD, Hussain AS. Predicting human oral bioavailability using in silico models. In: Krishna R, ed. Applications of Pharmacokinetic Principles in Drug Development. New York: Kluwer Academic/Plenum Publishers, 2004:53–74.
55. Grass GM, Sinko PJ. Physiologically-based pharmacokinetic simulation modeling. Advanced Drug Delivery Reviews 2002; 54:433–451.
56. Leung H. Physiologically-based pharmacokinetic modelling. In: Ballantyne, Marrs, Syversen, eds. General and Applied Toxicology. 2nd ed. New York, NY: Grove's Dictionarys, Inc(Macmillan Ltd.), 1999:141–154.
57. Gerlowski LE, Jain RK. Physiologically based pharmacokinetic modeling: principles and applications. J Pharm Sci 1993; 72(10):1103–1126.
58. Artursson P, Palm K, Luthman K. Caco-2 monolayers in experimental and theoretical predictions of drug transport. Advanced Drug Delivery Reviews 1996; 22:67–84.
59. Yee S. In vitro permeability across caco-2 cells (colonic) can predict in vivo (small intestinal) absorption in man—myth or fact? Pharm Res 1997; 14(6):763–766.
60. Florence AT. The oral absorption of micro- and nanoparticles: neither exceptional nor unusual. Pharm Res 1997; 14(3):259–266.
61. Park K, ed. Controlled Drug Delivery: Challenges and Strategies. Washington, DC: American Chemical Society, 1997.
62. Fan LT, Singh SK. Controlled Release. Berlin, Germany: Springer Verlag, 1989.
63. Porter CJH. Drug delivery to the lymphatic system. Critical Reviews[TM] in Therapeutic Drug Carrier Systems 1997; 14(4):333–393.
64. Jandacek RJ. The effect of nonabsorbable lipids on the intestinal absorption of lipophiles. Drug Metab Rev 1982; 13(4):695–714.
65. Noyes AA, Whitney WR. The rate of solution of solid substances in their own solutions. J Am Chem Soc 1897; 19:930–934.
66. Higuchi WI, Hiestand EN. Dissolution rates of finely divided drug powders I: Effect of a distribution of particle sizes in a diffusion-controlled process. J Pharm Sci 1963; 52(1):67–71.
67. Dressman JB, Fleisher D. Mixing-tank model for predicting dissolution rate control of oral absorption. J Pharm Sci 1986; 75(2):109–116.

68. Oh DM, Curl RL, Amidon GL. Estimating the fraction dose absorbed from suspensions of poorly soluble compounds in humans. Pharm Res 1993; 10(2):264–270.

69. Ebel JP, Jay M, Beihn RM. An in vitro/in vivo correlation for the disintegration and onset of drug release from enteric-coated pellets. Pharm Res 1993; 10(2):233–238.

70. Higuchi T. Rate of release of medicaments from ointment bases containing drugs in suspension. J Pharm Sci 1961; 50(10):874–875.

71. Narasimha B, Peppas NA. The role of modeling studies in the development of future controlled-release devices. In: Park K, ed. Controlled Drug Delivery: Challenges and Strategies. Washington, DC: American Chemical Society, 1997.

72. FDA Dissolution Testing of Immediate Release Solid Oral Dosage Forms. Washington, DC: Food and Drug Administration, Center for Drug Evaluation and Research, 1997.

73. Amidon GL, Lenneras H, Shah VP, Crison JR. A theoretical basis for a biopharmaceutic drug classification: the correlation of in vitro drug product dissolution and in vivo bioavailability. Pharm Res 1995; 12:413–420.

74. Yu LX. An integrated model for determining causes of poor oral drug absorption. Pharm Res 1999; 16(12):1883–1887.

75. Oberle RL, Amidon GL. The influence of variable gastric emptying and intestinal transit rates on the plasma level curve of cimetidine; an explanation for the double peak phenomenon. J Pharmacokin Biopharm 1987; 15(5):529–544.

76. Price JM, Biava CG, Oser BL, Vogin EE, Steinfeld J, Ley HL. Bladder tumors in rats fed cyclohexylamine or high doses of a mixture of cyclamate and saccharin. Science 1970; 167:1131–1132.

77. Kroes R, Peters PWJ, Berkvens JM, Verschuuren HG, De Vreis T, VanEsch GJ. Long-term toxicity and reproduction study with cyclamate, saccharin and cyclohexylamine. Toxicology 1977; 8:285–300.

78. Drasar BS, Renwick AG, Williams RT. The role of the gut flora in the metabolism of cyclamate. Biochem J 1972; 129:881–890.

79. Eichelbaum M, Hengstmann JH, Rost HD, Brecht T, Dingler HJ. Pharmacokinetics, cardiovascular effects, and metabolic actions of cyclohexylamine in man. Arch Toxikol 1974; 31:243–263.

80. Litchfield MH, Swan AAB. Cyclohexylamine production and physiological measurements in subjects ingesting sodium cyclamate. Toxicol Appl Pharmacol 1971; 18:535–541.

81. Buss NE, Renwick AG, Donaldson KM, George CF. The metabolism of cyclamate to cyclohexylamine and its cardiovascular consequences in humans. Toxicol Appl Pharmacol 1992; 115:199–210.

82. Mancia G, Ferrari A, Gregorini L, et al. Blood pressure and heart rate variabilities in normotensive and hypertensive human beings. Circ Res 1983; 53:96–104.

83. Kuipers F, Havinga R, Bosschieter H, Toorop GP, Hindriks FR, Vonk RJ. Enterohepatic recirculation in the rat. Gastroenterology 1985; 88:403–411.

84. Frederick CB, Potter DW, Chang-Mateu MI, Andersen ME. A physiologically based pharmacokinetic and pharmacodynamic model to describe the oral dos-

ing of rats with ethyl acrylate and its implications for risk assessment. Toxicol Appl Pharmacol 1992; 114(2):246–260.

85. Fiserova-Bergerova V, Diaz ML. Determination and prediction of tissue-gas partition coefficients. Int Arch Occup Environ Health 1986; 58:75–87.

86. Poulin P, Krishnan K. A tissue composition-based algorithm for predicting tissue: air partition coefficients of organic chemicals. Toxicol Appl Pharmacol 1996; 136:126–130.

87. Bronaugh RL, Stewart RF, Congdon ER, Giles AL. Methods for percutaneous absorption studies. I. Comparison with in vivo results. Toxicol Appl Pharmacol 1982; 62:474–480.

88. Scheuplein RJ, Bronaugh RL. Percutaneous absorption, In Lowell A. Goldsmith ed. Vol 2. Biochemistry and Physiology of the Skin. Oxford University Press, 1983:1255–1295.

89. Spreafico F, Zuccato E, Marcucci M, Sironi M, Paglialunga S, Madonna M, Mussinin E. Pharmacokinetics of ethylene dichloride in rats treated by different routes and its long-term inhalatory toxicity. In: Arnes B, Infante P, Reitz R, eds. Banbury Report 5, Ethylene Dichloride: A Health Risk?. Cold Spring Harbor, NY: Cold Spring Harbor Laboratory, 1980:107–129.

90. Reitz RH, Fox TR, Ramsey JC, Quast JC, Langvardt PW, Watanabe PG. Pharmacokinetics and macromolecular interactions of ethylene dichloride in rats after inhalation or gavage. Toxicol Appl Pharmacol 1982; 62:190–204.

91. D'Souza RW, Francis WR, Andersen ME. Physiological model for tissue glutathione depletion and increased resynthesis after ethylene dichloride exposure. J Pharmacol Exp Therap 1988; 245(2):563–568.

92. D'Souza RW, Andersen ME. Physiologically based pharmacokinetic model for vinylidene chloride. Toxicol Appl Pharmacol 1988; 95:230–240.

93. Gargas ML, Clewell HJ, Andersen ME. Metabolism of dihalomethanes in vivo: differentiation of kinetics constants for two independent pathways. Toxicol Appl Pharmacol 1986; 82:211–223.

94. Andersen ME, Gargas ML, Jones RA, Jenkins LJ. Determination of the kinetic constants of metabolism of inhaled toxicants in vivo based on gas uptake measurements. Toxicol Appl Pharmacol 1980; 54:100–116.

95. Ramsey JC, Andersen ME. A physiologically based description of the inhalation pharmacokinetics of styrene in rats and humans. Toxicol Appl Pharmacol 1984; 73:159–175.

9

Safety of Pesticidal Proteins in Food

Chris A. Wozniak

National Program Leader for Food Biotechnology and Microbiology, USDA-CSREES-PAS/CP, Washington, D.C., U.S.A.

John L. Kough

Biopesticides and Pollution Prevention Division, U.S. EPA, Office of Pesticide Programs, Washington, D.C., U.S.A.

INTRODUCTION AND REGULATORY BACKGROUND

Plants naturally produce numerous proteins and other compounds that deter or destroy pests and diseases. Although these protective components have always been present in plants, found in the food supply, and consumed by humans for many generations at some level in the diet, their induction and function are only now beginning to be understood by plant researchers. In addition, plants have recently been subjected to genetic manipulation involving the direct introduction of new protective traits that had never been a part of the food plant's genetic complement. Both these types of protective traits, naturally occurring and introduced by genetic engineering, would be considered pesticidal under the pesticide definition found in U.S. law. The U.S. Environmental Protection Agency (EPA) has determined that the naturally occurring pesticidal protein present in all crop plants need not be regulated. They are specifically exempted from registration requirements because their safe use in food and feed, without causing side effects, has a considerable history.

A pesticide is defined as a substance or mixture of substances intended to prevent, destroy, repel, or mitigate a pest. The pesticide definition also includes plant growth regulators, defoliants, and desiccants, as well as nitrogen stabilizers. The plant-expressed proteins that are discussed in this chapter, while clearly intended to control pests, cannot be readily placed in these categories by the public because they are accustomed to considering pesticides analogous to synthetic chemicals.

Pesticides are regulated in the United States, and U.S. possessions and territories by the U.S. EPA under the authority of the Federal Insecticide, Fungicide and Rodenticide Act (FIFRA) and the Federal Food, Drug and Cosmetic Act (FFDCA). FIFRA requires the EPA to determine if the pesticide's use presents any unreasonable adverse effects on the environment. Under the FFDCA, EPA must determine the human dietary risk from residues that result from use of a pesticide when that pesticide is applied to food or feed crops. EPA is required by the FFDCA to set a safe maximum residue level (food tolerance) for a pesticide applied to food or feed (1). The safety standard for pesticide residues under FFDCA is the reasonable certainty that no harm will result from the aggregate exposure to the pesticide residues. When a pesticide is shown to be safe regardless of the anticipated residue level, EPA can approve an exemption from the requirement for a food tolerance, meaning there is no set limit for this substance remaining on the crop or products derived from it in commerce.

While conventional synthetic chemical pesticides have been used for decades on crops and are what most people recognize as pesticides, the broad definition of a pesticide in FIFRA Section 2(u) gives a wider scope than is usually realized. Microbial agents, naturally occurring biochemicals, and pesticidal substances expressed in plants are all included within FIFRA's purview as pesticides. As would be expected, these types of pest control agents raise very different safety questions compared to conventional pesticides. EPA has constructed a different method for analyzing the risks for these biologically based pest control systems considering both their natural occurrence and their previous human exposure in the environment and the diet. Much of this guidance for safety testing of biological compounds is found in publicly available fora (2–4).

The pesticidal substances expressed in plants, originally referred to as "Plant Pesticides," are now called "plant-incorporated protectants" (PIPs) and are further described in the PIP rule (5). The PIP actually includes not only the pesticidal substance (i.e., the product of the genes introduced into the plant) but also the genetic material necessary for their production in planta. The PIP definition includes the genetic material necessary for the production of the pesticidal substance because FIFRA is an intent-based act. If a person intends to produce a substance for preventing, destroying, repelling, or mitigating a pest, that person is producing a pesticide. EPA determined in the PIP rule that the introduction of the genetic material coding for a pesticidal substance was an act indicating intent to produce a pesticide. However, it should be stressed that the risk assessment for environmental and dietary safety is based on the toxicity and exposure to the expressed pesticidal substance. In some instances, the movement of the genetic material to other sexually compatible plants will be part of the safety assessment, because this movement of the genetic material may lead to unexpected environmental exposures to the pesticidal substance itself.

Several PIPs with the δ-endotoxins of *Bacillus thuringiensis* (*Bt*) have been registered by the EPA, being the most widely adapted for use in plants. These include δ-endotoxins from a variety of strains of *Bt*, such as Cry1Ac in cotton, Cry1Ab in corn, and Cry 3A in potato, which have differing specificities toward the insect pests they are intended to control (6,7). It is important to note that the pesticidal substance and the genetic material responsible for its production are the registered PIP, not the plant that is producing the pesticidal trait. This distinction can have important ramifications with respect to pesticide-labeling issues, plant-breeding methodologies for PIP development, and experimental use permits for field testing, but will not be covered in detail herein.

In addition to the δ-endotoxins of *Bt* introduced into corn, cotton, and potato as registered PIPs, a replicase gene bestowing resistance to potato leaf roll virus has also been approved for use in potato. The majority of the registered PIPs also utilize selectable marker genes linked to the PIP trait. Marker genes encode proteins with enzymatic activity that easily identifies transformed cells in the background of normal plant cells in tissue culture (8,9). Most PIP traits are not easily detected directly so a marker gene allows one to select for plant cells in culture, which integrated the new genetic material. While the introduced traits phenotypically vary greatly, they do share a common biochemical essence, which simplifies the dietary risk assessment. All of the PIPs seen to date are proteins. Proteins consist of a linear series of amino acid residues linked by peptide bonds, assume a three-dimensional folding pattern dictated by their secondary structure and environment, and are generally susceptible to one or more digestive enzymes as are typically found in mammalian monogastric digestive systems.

The analysis of PIPs prior to their approval for commercial use and entrance into the food and feed supply is unprecedented in the field of food safety risk assessment. Plants expressing pesticidal substances such as the *Bt*-derived proteins are scrutinized more carefully than their conventionally bred counterparts. A key consideration in the evaluation of foodstuffs is the precision with which genetic modification through engineering of recombinant DNA modifies existing crop varieties as compared with the less precise mixing of thousands of genes as mediated by conventional plant breeding, which may include numerous intended but undescribed mutations induced by radiation or chemical mutagenesis (8).

DIETARY FATE OF PROTEINS

The essential nature of proteins as part of a healthy mammalian diet has been demonstrated and described in numerous fora (10,11). Dietary requirements for nitrogen and amino acids residues as metabolic building blocks have made proteins a key in the formulation of nutritional supplements and a mainstay of food production. The primary digestive enzymes of the mammalian monogastric stomach and small intestine, pepsin and trypsin/chymotrypsin, respectively, have evolved to deal with various proteins and glycoproteins such that they can be rendered useful for later absorption by the body. In this context, these proteins from animal, plant, and microbial sources differ considerably from many conventional pesticides that often comprise synthetic molecules without significant history of contact or consumption by animals.

Conventional chemical pesticides are often complex, heterocyclic ring structures with side-chain moieties that are not naturally found in the environment. Many chemical pesticides have direct acute toxicity or chronic effects and some even suggest reproductive effects in animal test systems. These uncertain metabolic fates and effects distinguish conventional pesticides from naturally occurring protein molecules that could be predictably digested to smaller subcomponents or residues by enzymatic action in the digestive tract. The uncertainties demonstrated in test systems also complicate the safety assessment of conventional pesticides compared with that of protein PIPs. As such, these newer generation pesticides (i.e., PIPs), while sharing a legal requirement for regulation by FIFRA and FFDCA, are biologically worlds apart from the more commonly used organophosphate, carbamate, and pyrethroid insecticides.

Proteins are known to first undergo significant digestion in the highly acidic lumen of the stomach, largely through the presence of secreted pepsin. The lability of the proteins in this environment will vary based upon amino acid sequence, presence of other food or buffering agents, health of the individual, and length of residence time in the stomach. Following passage from the stomach into the duodenum and the requisite rise in pH, other enzymes including chymotrypsin and trypsin are allowed to digest the protein-containing chyme. Similarly, the subsequent protein degradation rate in the small intestine will be influenced by factors that affect the pH of the intestinal contents, presence of buffering agents, secretion of pancreatin, and the individual characteristics of the protein. Proteins and peptide fragments will continue to be processed as they move into the jejunum and ileum with subsequent absorption via the circulatory system along the course of the small intestine.

Absorbed amino acids and small peptides will form the basis of catabolic syntheses of new proteins and be used as a source of substrate for sugar synthesis depending on nutrient availability at the time of absorption and circulation. Assimilation of amino acids into new protoplasm or deamination in the liver, with subsequent formation of urea, is a routine function for which mammalian systems have evolved precise controls. Uptake by specific transporters in the microvillar membranes and cytoplasm of intestinal epithelial cells ensures the fate of digested proteins. Against the background of the normal catabolic fate for proteins—degradation, absorption, and reuse—there are some real albeit few hazards with proteins. The primary issues are direct toxicity and allergenicity.

TOXICITY OF PROTEINS

In instances where proteins have mammalian toxicity, they have been shown to act acutely and generally at low doses (12). Bacterial toxins, for example, are known to work as single- or two-component systems and bind to cell receptor molecules, enter the cell or provide for the entry of another toxin component, and provide an avenue for the toxic protein to exact its destruction through various types of catalytic activity. The specific receptors on the microvilli of insect gut epithelium, which form the brush-border membranes, ensure the fate of insecticidal proteins targeting susceptible species. A vast milieu of proteinaceous toxins from microbial sources exist, which attack eukaryotic cells, such as pore formers, proteases, glucosidases, deamidases, ADP ribosylase, nucleases, excitotoxins, and N-glycosidases (ribosomal inhibitory proteins, RIPs) (13–19). The latter group, RIPs, may act in one- or two-component mechanisms, but do so in amounts that reflect their high potency: IC_{50}s are often < 1 nM and LC_{50}s for mice of 1 to 80 μg/kg body weight for type 2 RIPs (20). As with most other proteinaceous toxins from this list, their action as catalysts does not require significant amounts of substance to effect a morbid pathology in the host tissue. Some toxins are virulence factors that may be essential for pathogenicity of the microbial producer but not as devastating without the pathogen being present.

Other proteins may act as antinutrients and antidigestives or interfere with normal mammalian metabolism in less cell-specific manners. These proteins effect a chronic pathology that is less likely to be observed as an acute toxicological response, especially when consumed as part of a complex diet. Protease inhibitors, lectins, α-amylase inhibitors, hemi-lectins, arcelins, and others, when ingested, may influence digestion of proteins and starches, nutrient availability, and direct toxicity (e.g., inhibition of protein synthesis) on specific cell types and subgroups (21–24).

If any protein from the above-listed toxins were to be suggested as pesticidal substances in a food or feed crop, the level of scrutiny and toxicity testing required would be substantially greater than is required for proteins without a documented history of mammalian toxicity. Certainly, it is doubtful that proteins with similarity to antinutrients, toxins, or mammalian enzyme inhibitors could gain approval as components of a PIP without extensive longer-term tests. As with all biological pesticides reviewed for potential use, the risk assessment is based on a combination of hazard and exposure assessments. General guidance on the toxicity of proteins and how that relates to novel proteins was treated in an EPA Scientific Advisory Panel (25).

Bt as a Microbial Insecticide

Bt-derived insecticides have been registered for use on crops on farms and in home gardens since 1961 in the United States, and have been in use elsewhere in the world for even longer (26). A naturally occurring pathogen of the silkmoth and other insects, *Bt* was discovered and characterized in Japan almost 100 years ago. Formulations of various strains of *Bt* have been developed for insect pest management, primarily for use on lepidopteran insects (e.g., cabbage looper, diamondback moth, armyworms); however, strains have also been selected for their insecticidal activity towards dipteran (e.g., blackflies, mosquitoes), hymenopteran (pine sawfly), and coleopteran (e.g., corn rootworm and Colorado potato beetle) insects as well (27–31). Individual isolates with activities across the taxonomic borders of insect orders have also been described, as well as outside of the class Insecta (e.g., bacteria, protozoa, and anematoda) (32–35).

Commercial formulations of *Bt* insecticides may contain insecticidal crystal proteins (ICPs) and other proteinaceous toxins, endospores, cell wall fragments, fermentation solids, and a variety of inert ingredients, which aid in dispersal, stability, and application of the product. The gene(s) encoding the primary component of the insecticidal activity in this bacterium has even been transferred to another bacterium, *Pseudomonas fluorescens*, for production and accumulation (36). This surrogate host bacterium was then heat-killed to produce a potent, UV-protected insecticidal toxin for delivery to crop pests (37). Combinations of toxins from *Bt* and *B. sphaericus* have also been created for mosquito control following introduction into *Enterobacter amnigenus* (38).

The principal insecticidal toxins present in commercial formulations of microbially derived *Bt* cell preparations are the δ-endotoxins (39). These ICPs are typically formed as crystalline inclusion bodies or parasporal proteins at the time of bacterial endospore formation. Endospore formation is usually triggered by a reduction in available nutrients, accumulation of waste products, and possibly other biochemical signals not yet understood, during the fermentation of *Bt*. Only cells of *Bt* are known to produce protein toxins of this type, although insecticidal toxins with some strong similarities exist in other species (e.g., *B. sphaericus*) (40,41).

The toxicity of the δ-endotoxins or Cry proteins, as they are also known, lies in the presence of numerous receptor proteins on the microvilli of insect midguts (42). These receptors are known to bind δ-endotoxins with a great deal of specificity and affinity in susceptible insects (36,43). Conversely, the absence of the appropriate receptors in nontarget or resistant insects prevents the binding of Cry proteins and any subsequent pathology (44). It is this specificity that makes *Bt* microbial biopesticides and PIPs derived from native or altered *cry* genes very attractive to both producers of these products and users alike. With respect to potential environmental impact on nontarget insects and other invertebrates exposed to δ-endotox-ins, this lack of binding and effect in the vast majority of invertebrate species simplifies the risk assessment and reduces the required number of studies that may have to be submitted to regulatory authorities (45).

Strains of *Bt* have been isolated from a variety of environments and selected for their potential as biological insecticides. The majority of strains commercialized to date have insecticidal activity against lepidopteran larvae, although strains are also registered for use on coleopteran and dip-teran insect larvae. In most microbial preparations of *Bt* biopesticides, the δ-endotoxins are delivered in a protoxin form (120–130 kDa), which must be cleaved by insect gut proteases to be activated (55–65 kDa) (46). Addi-tionally, the alkaline nature of the midgut lumen is required for solubility of crystal proteins (28). The specificity inherent in these biopesticides is a result of the requirement for the precise type of gut receptor to interact with the ingested Cry protein and to bind with it with high affinity (27,47). To effect this disease state, the Cry protein molecules must intercalate into the cell membrane at the site of the appropriate receptor and form a channel or pore in an otherwise patent cell (48,49). In some instances, Cry proteins have been shown to work in concert with one molecule synergizing the activ-ity of another endotoxin type, while having little or no activity by itself (50). The multiple interactions with the gut receptors and lipid bilayer are not presently known in detail for these cooperatively acting toxins; however, the specificity demonstrated by these proteins suggests a similar requirement for binding.

With typical single toxin activity, pore formation is achieved through the integration of multiple monomers of δ-endotoxin protein into the apical membrane of midgut epithelial cells following binding to the brush-border membrane receptors (51). Molecular interaction between monomers of Cry protein is necessary for pore formation, but it is now evident that this organization and interaction occur within the cell membrane after insertion into the planar lipid bilayer (48). The β-barrel pore formed following oligo-merization is approximately 10 to 20 Δ in diameter and formed by the inter-action of four or six monomers. The resultant large conductance state caused by a loss of electrolytes and cell integrity ultimately leads to blebbing of cells and ulceration of the gut lining. If a sufficient quantity of δ-endo-toxin has bound to the brush-border microvilli receptors, feeding inhibition may commence in a matter of hours to a couple days. Typically, death will occur in two to five days, although the presence of *Bt* δ-endotoxins may also result in a feeding inhibition and larval death because of starvation in the absence of alternative food sources.

Nontarget toxicity unrelated to the prerequisite binding of a δ-endotoxin to a specific gut receptor protein in a susceptible insect has been documented with one variety of *Bt*. When alkali-solubilized toxins from a *Bt* var. *israelensis* strain were introduced into mice through intravenous or subcutaneous injection, paralysis and mortality ensued (52). Lysis of cultured mammalian cell lines and erythrocytes was also observed with this same toxin preparation. Per os dosing of mice did not, however, result in any indication of toxicity. Members of this subgroup of *Bt* most often demonstrate activity against dipteran larvae such as blackflies and mosquitoes.

While *Bt* var. *israelensis* strain IFC-1 exhibited the expected insecticidal activity against *Aedes aegypti* (yellow fever mosquito) larvae, it was also lethal to a host of other, unrelated insects when an alkali-solubilized toxin preparation was introduced by injection into the hemocoel (53). These same insects including *Trichoplusia ni* (cabbage looper), *Periplaneta americana* (American cockroach), *Oncopeltus fasciatus* (milkweed bug), *Heliothis zea* (corn earworm), and *Musca domestica* (house fly) were not affected by the alkali-solubilized toxin when administered by the oral route. Similarly, mice that were provided with this solubilized toxin or the crystalline preparation were not affected by an oral dose, but significant mortality was realized when intraperitoneal injection of the alkali-solubilized toxin was performed.

Interestingly, in this study, a parallel set of treatments with a *Bt* var. *kurstaki* did not yield toxicity following injection of the toxin into the body cavities of the insect species or the mouse test animals. Chromatographic analysis of the *Bt* var. *israelensis* crystal proteins indicated that there were several proteins in the range of 23.5 to 35 kDa based upon SDS-PAGE analysis. Removal of the 35-kDa peptide(s) did not result in a change in the relative toxicity of the preparation, although the remaining 23.5 to 24 kDa fraction and the 27 to 27.5 kDa fraction did have differing activities. While the neuromuscular symptoms observed with injection into *T. ni* were evident with both fractions, only the 27 to 27.5 kDa protein fraction maintained activity against *A. aegypti* (53).

A 28 kDa protein was found to be implicated in the mortality and cytotoxicity noted in mice and cultured mammalian cells following testing of solubilized crystal proteins of *Bt* var. *israelensis* (28,52,54–57). A protein moiety from the *Bt* var. *shandongiensis*, of similar relative molecular mass (i.e., 28 kDa) was also implicated as responsible for observed toxicity to cultured human leukemic (MOLT-4) cells (58). This class of δ-endotoxins, while incorporated in the crystal protein inclusion, does not share the normal specificity and requirement for a cell receptor to induce cellular leakage and lysis. They have been termed "Cyt toxins" for their general cytolytic abilities. While primarily restricted to strains of *Bt* var. *israelensis*, they have been found in at least three other subspecies or varieties (*jegathesan*, *medellin*, and *shandongensis*) (59–61). Similarly, parasporin, one of the inclusion proteins present in some *Bt* parasporal crystals, was found to recognize human leukemia cells. Other inclusion proteins with lectin-like activity have also been observed in some strains of *Bt* (62). None of these varieties has been a source of genetic material for development of pesticidal substances to be expressed in plants (i.e., PIPs).

The presence of ICP encoding (*cry*) genes on resident plasmids in *Bt* has allowed for relatively straightforward cloning of the active component for incorporation into other microbes and plants (63). Additionally, some of the δ-endotoxins have been modified structurally to enhance expression and to provide for codon preferences during optimization for expression and stability in an alternative organism. One advantage of cloning the genes encoding the δ-endotoxins and expressing them in an alternative organism such as other microbes or plants is the isolation of this toxin class from others that may be present in the organism. *Bt* strains may also produce numerous other toxins that are not necessarily so benign: *cyt* toxins, β-exotoxins, bacteriocins, enterotoxins, and vegetative insecticidal proteins in addition to the δ-endotoxins of primary interest (48,64–67). While some of these may add to the insecticidal potency of the microbial biopesticide, their presence is not always welcome. β-Exotoxins are nucleotide analogs with significant toxicity toward many organisms, including humans. For this reason, their presence is not allowed in commercial preparations of biopesticides in the United States and toxicity tests specifically designed to detect their presence are incorporated into the quality control measures instituted as part of the approved manufacturing process for commercial products intended for use in the United States (21,68,69).

SAFETY ASSESSMENT FOR TOXICITY

Because the proteins designated as PIPs are intended to control a pest, there is a presumption of toxicity to at least the pest species. Therefore, the safety assessment for protein PIPs seeks to confirm that the protein does not also demonstrate toxicity toward nontarget species including mammals. The basis of toxicity assessments is testing the PIP pesticidal substance on selected representative species. For environmental risk assessments, this includes a range of ecological indicator species including nontarget insects, fish, birds, and aquatic invertebrates. The selection of species for ecological effects depends to a certain extent on the expected environmental exposure. General guidance for ecological safety testing can be found under the microbial pesticide testing guidelines and in the reports from several Scientific Advisory Panels (2,5,70–72).

Through selective placement of the crystal protein gene (*cry*) into plants, a more defined insecticidal entity can be extracted from its background milieu and expressed at varying levels depending on the promoter sequences used to define the expression parameters. To date, the expression of one or two δ-endotoxin genes into crop plants has defined the majority of PIPs registered by the U.S. EPA (Table 1). The inherent specificity of these insecticidal proteins and the ability to modify expression levels in planta have led to the rise in their popularity among producers in the United States and several other countries (Fig. 1).

Table 1 Plant-Incorporated Protectants Registered by and Pending Before the U.S. EPA

Active ingredient	Crop	Major target pests	Registrant
Cry1Ab (Event Bt 11) YieldGard®	Maize, field	European corn borer, southwestern corn borer, corn earworm,	Syngenta
Cry1Ab (Event Bt 11) Attribute®	Maize, sweet	European corn borer, southwestern corn borer, corn earworm, fall armyworm	Syngenta
Cry1Ab (MON 810) YieldGard®	Maize, field	European corn borer, southwestern corn borer, corn earworm, southern cornstalk borer, fall armyworm	Monsanto Corporation
Cry1Ab/Cry3Bb1 YieldGard Plus®	Maize, field	European corn borer, southwestern corn borer, corn earworm, fall armyworm, western, northern and Mexican corn rootworm	Monsanto Corporation
Cry1F (TC1507) Herculex®	Maize, field	European corn borer, southwestern corn borer, corn earworm, fall armyworm, black cutworm	Dow/Mycogen Seeds and DuPont/Pioneer Hi-Bred International
Cry3Bb1 (MON 863) YieldGard Rootworm™	Maize, field	Western, northern, and Mexican corn rootworm	Monsanto Corporation
Cry 34Ab1/Cry35Ab1[a]	Maize, field	Western, northern, and Mexican corn rootworm	Dow/Mycogen Seeds and DuPont/Pioneer Hi-Bred International

Protein	Crop	Target pests	Company
Cry1Ac Bollgard®	Cotton, upland	Tobacco budworm, cotton bollworm, pink bollworm, beet and fall armyworm	Monsanto Corporation
Cry1Ac/Cry2Ab2 Bollgard II®	Cotton, upland	Tobacco budworm, cotton bollworm, pink bollworm, beet and fall armyworm	Monsanto Corporation
Cry1F/Cry1Ac Widestrike®[a]	Cotton, upland	Tobacco budworm, cotton bollworm, pink bollworm	Dow/Mycogen Seeds and DuPont/Pioneer Hi-Bred International
VIP3A (vegetative insecticidal protein 3A) VIP3A Cotton-102[a]	Cotton, upland	Tobacco budworm, corn earworm, fall and beet armyworm, black cutworm	Syngenta
Cry3A New Leaf®	Potato	Colorado potato beetle	Monsanto Corporation
Cry3A/Replicase (for potato leaf roll virus) New Leaf Plus®	Potato	Colorado potato beetle, potato leaf roll virus	Monsanto Corporation
Cry1Ac[a]	Tomato	Tomato pinworm, fruitworm and hornworm, cabbage looper	EHN Research

[a]Under review by the U.S. EPA and pending registration under FIFRA as of April, 2004.
Abbreviation: EPA, Environmental Protection Agency.

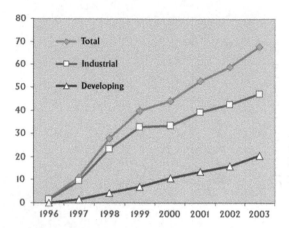

Figure 1 Global area of transgenic crops, 1996 to 2003 in industrial and developing countries (in millions of hectares). *Source*: From Ref. 73.

The major human exposure for PIP proteins in food plants is in the diet. The focus of mammalian testing of PIP proteins is therefore the oral route of exposure (25,74). An essential part of the risk assessment for pesticides depends on examining the hazards a substance presents and comparing its expected exposure in the diet. To adequately protect the public from harm because of the pesticide residues, EPA employs safety factors based on the results of toxicity testing in surrogate species and an expected exposure. Therefore, a high dose of test substance is given to the rodent test subjects in the oral toxicity tests. If no untoward effects are seen, then a comparison of the dosed amount to that likely to occur through consumption of plant materials provides a means to address the safety of the dietary exposure.

All new active ingredients with pesticidal intent are regulated under FIFRA and a series of toxicity test data and in-depth product characterization information are submitted to the Agency for review (75). Included within the required studies to be submitted by a registrant of a PIP is a maximum hazard dose, acute oral toxicity test, typically performed on mice or rats (70). This evaluation is based upon the lack of toxicity of most proteins and the observation that when proteins are acutely toxic, they generally induce a toxic response at a relatively low dose of material (12). The acute oral toxicity test performed as part of the EPA assessment of risk to human and animal health has a suggested dosing regime of 2000 to 5000 mg of test material per kilogram of body weight of the test animal.

The test material chosen for the mammalian toxicity evaluation of PIP proteins has typically been produced in a bacterial production system because of the large amount of pure protein required to perform the study. While the bacterial systems are more manageable and feasible for the production and purification process of adequate test amounts, the test material itself must still be verified as being equivalent to that which is produced in the plant. Differences in glycosylation and other potential translational modifications must be examined to ensure equivalency. A variety of biochemical and molecular characterizations are available to the registrant as a means of establishing this equivalency (e.g., definition of relative molecular mass through electrophoretic separation; serological detection using western blotting techniques or ELISA; mass spectrometry; MALDI-TOF/MS; ESI/MS; HPLC-ESI; terminal sequencing; X-ray crystallography; periodic acid–Schiff reagent staining, endo-H and other glycosyl residue detection techniques; and insect bioassay for potency).

The preferred test animal for the acute oral toxicity test is the rodent; all animals should be between 8 and 12 weeks old at the time of dosing. Weight variation among the animals must be less than 20% within either sex. At least five test animals of each sex are to be dosed with the test substance for comparison between the two groups; all female rodents should be nulliparous and not pregnant. Animals are fasted overnight prior to dosing. Subsequent to dosing, feed and water are made available ad libitum to the test animals during the observation period.

The test substance, consisting of the pesticidal substance and appropriate carrier (i.e., water or an aqueous dilute methylcellulose solution), is administered orally by gavage to rodent test animals. Volume of the dose should be limited to 2 mL/100 g body weight when administered in aqueous solution. Following dosing, the test animals are observed daily for any clinical or behavioral signs indicative of a toxic reaction to the pesticidal protein. Weight changes are recorded for comparison with pretest body weight values and data are reported both as summarized averages for the treatment group by sex and as an individual datum per animal. Any animal that dies during the course of the study is necropsied and all animals are euthanized at the end of the study and necropsied. If any animal is observed to be in distress or pain during the course of the observation period (14 days), the animal will be euthanized at that point and necropsied. Fourteen days is the most common observation period, but tests may extend to 28 days. Should any acute toxicity be noted in this limit test, then a subsequent test of longer duration and an altered dosing pattern (i.e., subchronic) will be required. Longer-term feeding studies have been carried out on some *Bt* proteins (76).

During the observation period following dosing, careful evaluation of the skin and fur, eyes and mucous membranes, respiratory and circulatory effects, autonomic effects such as salivation, central nervous system effects including tremors and convulsions, changes in the level of motor activity, gait, and posture, reactivity to handling or sensory stimuli, grip strength, and stereotypies or bizarre behavior (e.g., self-mutilation, walking backwards) must be performed and recorded. The individual weights of animals should be determined shortly before the test substance is administered, weekly thereafter, and at death. Changes in weights should be calculated and recorded when survival exceeds one day. Gross pathology is also recorded during the necropsy following the death of the animal or euthanization at the end of the test period.

Using the maximum hazard dose, limit test approach for evaluating acute oral toxicity of PIP and phenotypic marker proteins, no toxicity or acute reaction attributable to the ingestion of pesticidal proteins has been observed. When no effects are seen to be caused by the proteinaceous test substance even at relatively high dose levels in the acute oral exposure, the proteins are considered not toxic. The acute oral toxicity test has been performed in mice with a pure preparation of the PIP protein at doses ranging from 3280 to over 5000 mg/kg body weight (77). The high level of test material used reflects the desire to uncover any inherent toxicity in the material and greatly exceeds the range of expression levels of protein found in crop commodities grown under various conditions. The amounts of PIP proteins ingested in human or animal diet from consumption of crops directly or after processing is vastly exceeded by the dose administered in the acute oral toxicity test, when considered on a relative body weight basis.

In addition to the acute oral toxicity test submission, registrants of PIPs are required to submit a product characterization package, which details the biochemical and molecular nature of the pesticidal substance, as well as the method of introduction of the pesticidal trait. Included as part of this submission are a detailed restriction map of the introduced DNA, nucleotide sequence of the introduced DNA, amino acid sequence of the expressed traits, source of all nucleic acid elements (i.e., coding regions, promoters, termination sequences, introns, and modifying elements), function of the introduced nucleic acids, and copy number of the introduced gene cassette. Data supplied must also include an analysis of the transformation events, which indicates the integration and expression of the introduced DNA. This is typically accomplished by Southern blotting methodologies and PCR-based analyses for integrating data, and an ELISA-based expression analysis, which quantifies the presence of the pesticidal trait in various plant tissues and organs (e.g., roots, stems, leaves, pollen, nectar, grain, or fruit) as well as the whole plant average.

Proteins expressed as part of the integrated DNA are also evaluated for the presence of glycosylation. In some instances, glycosylated proteins may be immunogenically enhanced as compared to their nonglycosylated counterparts. As mentioned above, as part of the assessment of biochemical equivalency between proteins produced in planta versus those produced in a surrogate production system like *Escherichia coli*, glycosylation is also relevant to the production system because a protein produced in a prokaryotic system is far less likely to contain any added glycosyl residues. The presence or absence of glycosylation can be detected in several ways, including the use of agents to remove any possible glycosyl residues (e.g., endo-H and TFMS), followed by an assessment of gel mobility shift using SDS-PAGE. Staining of proteins separated by electrophoresis with reagents capable of detecting vicinal diols or other evidence of sugar residues (e.g., periodic acid–Schiff reagent and various proprietary kits) can also be applied to satisfy this data requirement. The key issue in examining the glycosylation of the test substance protein used in the oral toxicity test is the equivalency of the proteins and not the possible influence of glycosylation on allergenicity in general. In the δ-endotoxins examined to date, none have been found to be posttranslationally modified to contain glycosyl residues.

The mammalian intestinal mucosa lacks receptor target proteins specific for the δ-endotoxins of *Bt*. In contrast, susceptible insects are known to exhibit cadherin-like or aminopeptidase N type receptors on the surface of cells lining the midgut lumen. Histopathological examination of Brown Norway rat intestinal cells following consumption of Cry1Ab indicated no alterations when compared to control animal intestinal epithelia (78,79). An in vitro examination of Cry protein binding to intestinal tissues of rhesus monkeys, rats, mice, and humans was also performed and no evidence of binding detected. This lack of binding has correlated well with the observed absence of acute toxicity seen in rabbits, rats, and fish and in rodent studies submitted to the EPA for review (76,77,80,81).

Similarly, Hofman et al. (42) did not detect any binding of a δ-endotoxin from *Bt* var. *thuringiensis* to brush-border membrane vesicles (BBMV) from rat intestines. Vazquez-Padron et al. (82) did find some indication of Cry1Ac binding to the mucosal surface of mouse intestinal cells and to the BBMV prepared from them, which was accompanied by a transient hyperpolarization. In these experiments, the protoxin form of the δ-endotoxin was used to evaluate interactions with mouse intestinal sections and BBMV. Because this is not the form of the δ-endotoxin present in PIPs, and the protoxin is normally rapidly transformed in the insect gut by proteases, it is not clear how to interpret the relevance of these findings to dietary consumption of *Bt*-derived Cry proteins.

While the pesticidal substance expressed in planta and the nucleic acid necessary for its production are the regulated part of a PIP, a plant composition or proximate analysis comparison is occasionally part of the product characterization submitted to and reviewed by the Agency. In this comparison, the PIP-expressing plant in question and a nontransformed, near-isogenic line without the pesticidal substance may be contrasted in terms of protein, fatty acids, carbohydrate, minerals, acid ash, neutral ash, nitrogen, amino acids, grain, and oil components. Crop-specific constituents such as gossypol in cotton may also be examined. The intent of this analysis is to demonstrate the overall equivalency of the PIP plant to its conventional progenitor (83,84). This is accomplished by examining the key constituents of the plants and determining if they fall into the historical range for the crop species. This type of comparison is similar to that carried out by the company as part of their analysis for consultation with the Center for Food Safety and Applied Nutrition at the Food and Drug Administration.

SAFETY ASSESSMENT FOR FOOD ALLERGENICITY

Because the protective traits being discussed are proteins with expected dietary exposure, there is a unique aspect of the risk assessment for PIPs that differs from most other pesticide residues: potential food allergenicity. Food allergenicity could be considered as the ability of a protein present in the diet to incite food allergy in an individual. The condition of being allergic to food itself is unusual in that food items that are normally safely consumed by people can cause adverse reactions in sensitized individuals. The process of sensitization to a food component and the predisposition of an individual to develop food allergy are areas that could benefit from further research (85). However, it is clear that proteins or components that interact with proteins are responsible for triggering sensitization that leads to food allergy. Therefore, an assessment of the newly expressed protein in a PIP-expressing plant is necessary to address all aspects of dietary safety.

The assessment of a protein without prior known dietary exposure for potential food allergenicity is a new area of hazard assessment. The hazard here is the entry of a new food allergen into the food supply with the introduction of a novel protein into a food plant. This hazard must be considered against the background of the preexisting hazard from foodstuffs known to trigger allergy in sensitized individuals as well as from what is currently entering the food supply. New foods like kiwi fruit (*Actinidia chinensis*) were brought into the U.S. market and later found on broader consumption to induce allergies in certain consumers in the United States. A similar phenomenon has been noted with peanut (*Arachis hypogaea*) consumption and the incidence of peanut allergy in Europe. Also an interesting sidelight is the increase in some food allergies to fruits such as bananas and kiwis that are linked to the increased incidence of dermal latex allergy (86). In addition, it should be stressed that a wide range of proteins characterized less well than the traits being introduced into food plants through genetic engineering have been and are being consumed without untoward effects.

It is often stated that all food allergens are proteins. While this is correct, it is also true that many food proteins are consumed on a regular basis without adverse effects, by the entire human community including those with food allergies. The distinction between proteins that can be safely consumed and those that could trigger allergic responses is the crux of the food allergenicity assessment. Utilizing input from numerous scientific fora, the assessment of a novel protein in food for its potential allergenicity is based on a weight of the evidence approach wherein data from several sources are considered. The information includes a discussion of the source of the expressed trait, the allergenic history of the source organism, certain biochemical characteristics of any expressed novel protein, and the results of amino acid sequence similarity comparisons between the novel protein and databases for known allergenic proteins. The totality of this information on a new protein is examined, along with the likely dietary exposure, to determine if the new protein could present a risk as a new potential dietary allergen.

Companies provide information on many biochemical characteristics of the novel protein including the molecular weight, bioactivity, and similarity of the expressed protein compared to that found in the original source. While these characteristics are critical for the characterization of the protein, they are of lesser importance to the allergy assessment. The biochemical characteristics considered cogent to a food allergy assessment include protein stability to acid, heat, or protease degradation, and protein modifications like glycosylation (87–89). Molecular weight, the presence of posttranslational modifications, and antibody recognition are determined by standard methods based on differential membrane separation and staining. An essential aspect of these assays is a consideration of both the sensitivity and specificity of the assay for the analyte as well as inclusion of appropriate controls. Protein stability to acid, heat, and protease degradation are determined by methods that are not standardized like the other assays. Part of the difficulty in assessing these protein features is that they do not lend themselves to explicit endpoints like the presence or absence of carbohydrate residues and that their endpoints can be influenced by the detection system used in the assay.

It is critical to know the limitations of the assay system when judging the results for protein stability determinations. Some proteins bind the dyes used in gel staining to a lesser degree than others, lowering the sensitivity of detection and implying lower protein stability in this case. When using antibody-based reagents, a monoclonal antibody may recognize the disappearance of a single epitope whereas a polyclonal reagent would track a greater region of the protein, providing more information on the loss of protein integrity. The loss of bioactivity may be a cogent for efficacy determinations but not relevant to loss of antigenicity for the potential allergenicity question. The assay system may also be biased by the choice of test conditions increasing the concentrations of protease reagents to make a protein appear less stable. It may be more appropriate to examine the issue of protein stability to proteolysis as a kinetic phenomenon rather than a detection endpoint for the protein itself. A kinetic approach, while difficult in terms of detecting the products of protease enzymatic activity, may offer a means to eliminate the issue of detection method sensitivity driving the endpoint (90). Much effort has recently been aimed at validation and standardization of techniques that assay protein stability (91).

A critical component of the allergenicity assessment is the amino acid sequence similarity comparison between the expressed protein and a database of known allergenic proteins. The comparison is done by two mechanisms: an overall similarity comparison for the entire expressed protein and a stepwise comparison of six to eight amino acid fragments. A significant amino acid similarity of a novel protein to a known allergen for the overall comparison is a 35% amino acid identity over an 80% amino acid portion of the protein. There are examples of proteins that could fit this similarity criterion without being allergens, for example, the tropomyosins from crustaceans being related to those from mammals. It is also clear that there is no accounting at present for the repeated presence of certain amino acid sequences that provide a structural motif for proteins (e.g., beta-pleated sheets or alpha helices) without signaling an allergenicity hazard.

There is also considerable controversy on the use of a six versus an eight amino acid sequence of the novel protein as the basis for the stepwise sequence comparison. The concern is that there will be too many false positive similarities appearing if a six amino acid window is used and too many false negatives if an eight amino acid window is employed. The basis for the use of the stepwise window approach is the belief that a contiguous eight amino acid epitope is the minimum needed for cross-linking two IgE molecules on the surface of a mast cell to trigger degranulation. IgE cross-linking and degranulation is the first step in the type I hypersensitivity reaction that is linked to food allergy.

The possibility that an epitope is nonlinear and presents as a discontinuous epitope is part of the rationale for utilizing less than an eight amino acid window. There is evidence that the six amino acid comparison will yield false positives in that typical constituent proteins not known to be allergenic are indicated as possible allergens using the six amino acid similarity screen (92). No sequence analysis method has yet been conceived that can identify discontinuous epitopes or could account for the hazard of discontinuous epitopes being present in novel proteins. There is an additional method for screening the positive results from the six amino acid identity search for the potential antigenicity. Using the hydrophobicity or hydrophilicity rating of the six amino acid match and the known epitope maps of allergens, several algorithms can predict the antigenicity of the predicted match and need for serum screens and further studies (93). This system may refine the six-amino acid screen by reducing the number of false positives requiring further study. However, there is still no indication that the difference between the six versus the eight amino acid method represents an enhanced level of detecting new potential allergens. No novel protein that has passed the eight-amino acid screen has shown to be responsible for a new reported human allergy.

Taken as a part of a weight-of-evidence approach, all the screening methods discussed above can give an indication of potential allergenicity. However, no method that has been suggested for further studying a protein identified as a potential allergen has been validated. This includes the targeted serum screen and animal models of food allergy. The identification, collection, performance quality control, and maintenance of a bank of hypersensitive patient sera for even the major food allergens is a daunting task that has yet to be addressed. The use of these sera in a test format with appropriate controls for nonspecific binding and antibody titers that would indicate a positive recognition are also undefined.

Complicating this discussion is the fact that there are no validated animal model systems to mimic the development of food allergy (94–96). For traditional chemical pesticides and pharmaceuticals, there is a regular battery of safety tests utilizing animal surrogates for estimating human hazards. These tests are based on using an animal population to estimate the human response to an endpoint like toxicity or carcinogenicity, applying appropriate safety factors and estimating if the aggregate exposure to the test substance has a reasonable certainty of causing no harm. No validated animal models are available for the estimation of a food allergy hazard although there is currently an effort to develop tests to that end (85). The envisioned animal model for food allergy should respond like the most sensitive human subgroup, a population identified with atopy and dermatitis. Even this identified sensitive subgroup has a range of distinct proteins to which they become sensitized. No atopic individual has become sensitized to all the proteins identified as food allergens. Given these conditions, it is difficult to imagine a model system that could mimic this human response pattern. However, having an available model test system would facilitate the safety determination if the indications from the existing screening procedure are equivocal.

CONCLUSIONS

The assessment of dietary safety for novel proteins being introduced into food plants as PIPs provides a reasonable certainty that no harm will result from the aggregate exposure to that PIP protein. The basis for this assessment is that the biochemical nature of proteins provides a certain predictable dietary fate to ingested proteins: breakdown to peptides and single amino acids that can be shown to present few hazards. The major issues relating to PIP proteins are their acute toxicity, because they are demonstrated to control pests, and their potential to incite allergic responses in sensitized individuals. The toxicity is determined by an acute oral toxicity test done with pure PIP protein at doses high enough to ensure safe consumption at expected dietary exposure levels. The allergenic potential is assessed by amino acid similarity comparisons to known allergens and examination of biochemical characteristics that would suggest that the protein may not behave in a similar manner to typical dietary proteins. Further testing may be warranted if there are significant adverse effects in the acute toxicity test or indications that the PIP protein has a similarity to known toxins, antinutrients, or allergens.

DISCLAIMER

This chapter has been reviewed by the Environmental Protection Agency's Office of Pesticide Programs and approved for publication. Approval does not signify that the contents necessarily reflect the views and policies of the Agency, nor does mention of trade names or commercial products constitute endorsement or recommendation.

REFERENCES

1. National Research Council. Strengths and weaknesses of the current regulatory framework. Genetically Modified Pest-Protected Plants: Science and Regulation. Washington, D.C.: National Academy Press, 2000:149–151.
2. EPA. OPPTS. Harmonized Guidelines, Series 885 Microbial Pesticide Test Guidelines—Final Guidelines, 2004. (Accessed 02/25/04. http://www.epa.gov/opptsfrs/OPPTS_Harmonized/885_Microbial_Pesticide_Test_Guidelines/Series/index.html).
3. OSTP. MON 810 Maize Case Study No. II, Office of Science and Technology Policy, Council on Environmental Quality, Executive Office of the President, 2001 (http://www.ostp.gov/html/ceq_ostp_study3.pdf).
4. OECD. Safety evaluation of foods derived by modern biotechnology concepts and principles, 1993 (http://www.oecd.org/topic/0,2686,en_2649_34537_1_1_1_1_37437,00.html).
5. EPA. Regulations Under the Federal Insecticide, Fungicide and Rodenticide Act for Plant-incorporated Protectants (formerly Plant-Pesticides). Federal Register 2001; 66(139), Rules and Regulations, pp. 37772–37817.
6. EPA. Regulating Pesticides—Plant-incorporated Protectants; Currently Registered Section 3 PIP Registrations, 2004 (http://www.epa.gov/oppbppd1/biopesticides/pips/index.htm).
7. Crickmore N. *Bacillus thuringiensis* Toxin Nomenclature, 2004 (http://www.biols.susx.ac.uk/home/Neil_Crickmore/Bt/).
8. Donaldson L, May R. Health implications of genetically modified foods. UK Department of Health, 1999 (http://www.dh.gov.uk/assetRoot/04/06/50/90/04065090.pdf).
9. Fuchs RL, Ream JE, Hammond BG, Naylor MW, Leimgruber RM, Berberich SA. Safety assessment of the neomycin phosphotransferase II (NPT II) protein. Bio/Technology 1993; 11:1543–1547.
10. Organisation for Economic Co-operation and Development (OECD). Considerations for the Safety Assessment of Animal Feedstuffs Derived from Genetically Modified Plants, OECD, Paris, 2003.
11. Pellett PR, Young VR. Techniques for protein quality evaluation: background and discussion. In: Peter LP, Vernon RY, eds. Nutritional Evaluation of Protein Foods. Part I. Tokyo, Japan: United Nations University Press. ISBN 92-808-0129-5 (http://www.unu.edu/unupress/unupbooks/80129e/80129E00.htm).

12. Sjoblad RD, McClintock JT, Engler R. Toxicological considerations for protein components of biological pesticide products. Regul Toxicol Pharmacol 1992; 15:3–9.
13. Kuhnert P, Berthoud H, Straub R, Frey J. Host cell specificity of RTX toxins from haemolytic *Actinobacillus equuli* and *Actinobacillus suis*. Vet Microbiol 2003; 92:161–167.
14. Menestrina G, Serra MD, Prevost G. Mode of action of β-barrel pore-forming toxins of the staphyloccocal α-hemolysin family. Toxicon 2001; 29: 1661–1672.
15. Oh D, Shin SY, Lee S, et al. Role of the hinge region and the tryptophan residue in the synthetic antimicrobial peptides, Cecropin A91–8-Magainin 2(1–12) and its analogues, on their antibiotic activities and structures. Biochemistry 2000; 39:11855–11864.
16. Rossetto O, Tonello F, Montecucco C. Proteases. In: Burns DL, Barbieri JT, Iglewski BH, Rappuoli R, eds. Bacterial Protein Toxins. Washington, D.C.: ASM Press, 2003:271–282.
17. Kahn RA, Gilman AG. The protein cofactor necessary for ADP-ribosylation of G_s by cholera toxin is itself a GTP binding protein. J Biol Chem 1986; 261:7906–7911.
18. Yang DD, Kuan C-Y, Whitmarsh AJ, et al. Absence of excitotoxicity-induced apoptosis in the hippocampus of mice lacking the Jnk3 gene. Nature 1997; 389:865–870.
19. Bagga S, Hosur MV, Batra JK. Cytotoxicity of ribosome-inactivating protein saporin is not mediated through α2-macroglobulin receptor. FEBS Lett 2003; 541:16–20.
20. Melton-Celsa A, O'Brien AD. Plant and bacterial toxins as RNA *N*-glycosidases. In: Burns DL, Barbieri JT, Iglewski BH, Rappuoli R, eds. Bacterial Protein Toxins. Washington, D.C.: ASM Press, 2003:245–255.
21. Gutierrez-Campos R, Torres-Acosta JA, Saucedo-Arias LJ, Gomez-Lim MA. The use of cysteine proteinase inhibitors to engineer resistance against potyviruses in transgenic tobacco plants. Nat Biotechnol 1999; 17:1223–1226.
22. Reynoso-Camacho R, González de Mejia E, Loarca-PiZa G. Purification and acute toxicity of a lectin extracted from tepary bean (*Phaseolus acutifolius*). Food Chem Toxicol 2003; 41:21–27.
23. Carlini CR, Guimares JA. Plant and microbial toxic proteins as hemilectins: emphasis on canatoxin. Toxicon 1991; 29:791–806.
24. Carlini CR, Grossi-de Sa MF. Plant toxic proteins with insecticidal properties. A review on their potentialities as bioinsecticides. Toxicon 2002; 40:1515–1539.
25. EPA. Mammalian Toxicity Assessment Guidelines for Protein Plant Pesticides, Presentation to the FIFRA Scientific Advisory Panel and Final Report, June 7, 2000. (Accessed February 5, 2004. http://www.epa.gov/oscpmont/sap/2000/june/mammaltox.pdf and http://www.epa.gov/oscpmont/sap/2000/june/finbtmamtox.pdf.).
26. Siegel J. The mammalian safety of *Bacillus thuringiensis*-based insecticides. J Inv Path 2001; 77:13–21.

27. Knowles BH, Thomas WE, Ellar DJ. Lectin-like binding of *Bacillus thuringiensis* var. *kurstaki* lepidopteran-specific toxin is an initial step in insecticidal action. FEBS Lett 1984; 168:197–202.
28. Siegel JP, Shadduck JA, Szabo J. Safety of the entomopathogen *Bacillus thuringiensis* var. *israelensis* for mammals. J Econ Entomol 1987; 80:717–723.
29. Garcia-Robles I, Sánchez J, Gruppe A, et al. Mode of action of *Bacillus thuringiensis* PS86Q3 strain in hymenopteran forest pests. Ins Biochem Mol Biol 2001; 31:849–856.
30. Oehme FW, Pickerell JA. Genetically engineered corn rootworm resistance: Potential for reduction in human health effects from pesticides. Biochem Environ Sci 2003; 16:17–28.
31. Lavrik PB, Bartnicki DE, Feldman J, et al. Safety assessment of potatoes resistant to Colorado potato beetle. In: Engel KH, Takeoka GR, Teranishsi R, eds. Genetically Modified Foods: Safety Issues. Washington, D.C.: ACS, 1995: 148–158.
32. Zhong C, Ellar DJ, Bishop A, Johnson C, Lin S, Hart ER. Characterization of a *Bacillus thuringiensis* δ-endotoxin which is toxic to insects in three orders. JIP 2000; 76:131–139.
33. Yudina TG, Konukhova AV, Revina LP, Kostina LI, Zalunin IA, Chestukhina GG. Antibacterial activity of Cry- and Cyt-proteins from *Bacillus thuringiensis* ssp. israelensis. Can J Microbiol/Rev Can Microbiol 2003; 49:37–44.
34. Feitlson JS, Payne J, Kim L. *Bacillus thuringiensis*: insects and beyond. Bio/ Technology 1992; 10:271–275.
35. Wei J-Z, Hale K, Carta L, Platzer E, Wong C, Fang S-C, Aroian RV. *Bacillus thuringiensis* crystal proteins that target nematodes. Proc Natl Acad Sci USA 2003; 100:2760–2765.
36. Schnepf E, Crickmore N, Van Rie J, et al. *Bacillus thuringiensis* and its pesticidal crystal proteins. Microbiol Mol Bio Rev 1998; 62:775–806.
37. EPA. Fact Sheet: *Bacillus thuringiensis* delta endotoxins encapsulated in killed *Pseudomonas fluorescens* (006409, 006410, 006457, 006462), 2004 (http:// www.epa.gov/oppbppd1/biopesticides/ingredients/factsheets/fact-sheet_006409.htm).
38. Tanapongpipat S, Nantapong N, Cole J, Panyim S. Stable integration of mosquito-larvacidal genes from *Bacillus thuringiensis* subsp. *israelensis* and *Bacillus sphaericus* into the chromosome of *Enterobacter amnigenus*: a potential breakthrough in mosquito biocontrol. FEBS Microbiol Lett 2003; 221:243–248.
39. McClintock JT, Schaffer CR, Sjoblad RD. A comparative review of the mammalian toxicity of *Bacillus thuringiensis*-based pesticides. Pest Sci 1995; 45: 95–105.
40. Tanapongpipat S, Luxananil P, Promdonkoy B, Chewawiwat N, Audtho M, Panyim S. A plasmid encoding a combination of mosquito-larvacidal genes from *Bacillus thuringiensis* subsp. *israelensis* and *Bacillus sphaericus* confers toxicity against a broad range of mosquito larvae when expressed in Gram-negative bacteria. FEMS Microbiol Lett 2003; 228:259–263.
41. Siegel JP, Shadduck JA. Clearance of *Bacillus sphaericus* and *Bacillus thuringiensis* ssp. *israelensis* from mammals. J Econ Entomol 1990; 83:347–355.

42. Hofman C, Luthy P, Hutter R, Pliska V. Binding of the delta endotoxin from *Bacillus thuringiensis* to brush-border membrane vesicles of the cabbage butterfly (*Pieris brassicae*). Eur J Biochem 1988; 173:85–91.

43. English L, Slatin SL. Mode of action of delta-endotoxin from *Bacillus thuringiensis*: a comparison with other bacterial toxins. Insect Biochem Mol Biol 1992; 22:1–7.

44. Ballester V, Granero F, Tabashnik BE, Malvar T, Ferré J. Integrative model for binding of *Bacillus thuringiensis* toxins in susceptible and resistant larvae of the diamondback moth (*Plutella xylostella*). Appl Environ Microbiol 1999; 65:1413–1419.

45. EPA. Corn Rootworm Plant Incorporated Protectant Insect Resistance Management and Non-Target Insect Issues, Presentations to the FIFRA Scientific Advisory Panel and Final Report, August 27–19, 2002; accessed February 5, 2004(http://www.epa.gov/oscpmont/sap/2002/august/august2002final.pdf).

46. Mohan M, Gujar GT. Characterization and comparison of midgut proteases of *Bacillus thuringiensis* susceptible and resistant diamondback moth (*Plutellidae*: *Lepidoptera*). J Inv Path 2003; 82:1–11.

47. Hua G, Masson L, Jurat-Fuentes JL, Schwab G, Adang MJ. Binding analyses of *Bacillus thuringiensis* Cry δ-endotoxin using brush border membrane vesicles of *Ostrinia nubilalis*. Appl Environ Microbiol 2001; 67:872–879.

48. Promdonkoy B, Ellar DJ. Investigation of the pore-forming mechanism of a cytolytic δ-endotoxin from *Bacillus thuringiensis*. Biochem J 2003; 374:255–259.

49. Thomas WE, Ellar DJ. Mechanism of action of *Bacillus thuringiensis* var *israelensis* insecticidal δ-endotoxin. FEBS Lett 1983; 154:362–368.

50. Moellenbeck DJ, Peters ML, Bing JW, et al. Insecticidal proteins from *Bacillus thuringiensis* protect corn from corn rootworms. Nat Biotechnol 2001; 19: 668–672.

51. Soberón M, Prez RV, NuZez-Valdez ME, et al. Evidence for intermolecular interaction as a necessary step for pore-formation activity and toxicity of *Bacillus thuringiensis* Cry1Ab toxin. FEMS Microbiol Lett 2000; 191:221–225.

52. Thomas WE, Ellar DJ. *Bacillus thuringiensis* var. *israelensis* crystal delta-endotoxin effects on insect and mammalian cells in vitro and in vivo. J Cell Sci 1983; 60:181–197.

53. Cheung PYK, Roe RM, Hammock BD, Judson CL, Montague MA. The apparent in vivo neuromuscular effects of the δ-endotoxin of *Bacillus thuringiensis* var. *israelensis* in mice and insects of four orders. Pest Biochem Physiol 1985; 23:85–94.

54. Armstrong JL, Rohrmann GF, Beaudreau GS. Delta endotoxin of *Bacillus thuringiensis* subsp. *israelensis*. J Bacteriol 1985; 161:39–46.

55. Davidson EW, Yamamoto T. Isolation and assay of the toxic component from the crystals of *Bacillus thuringiensis* var. *israelensis*. Curr Microbiol 1984; 11:171–174.

56. Mayes ME, Held GA, Lau C, et al. Characterization of the mammalian toxicity of the crystal polypeptides of *Bacillus thuringiensis* subsp. *israelensis*. Fund Appl Toxicol 1989; 13:310–322.

57. Mizuki E, Ohba M, Akao T, Yamashita S, Saitoh H, Park YS. Unique activity associated with non-insecticidal *Bacillus thuringiensis* parasporal inclusions:

in vitro cell-killing action on human cancer cells. J Appl Microbiol 1999; 86: 477–486.

58. Lee D-W, Katayama H, Akao T, et al. A 28 kDa protein of the *Bacillus thuringiensis* serovar *shandongiensis* isolate 89-T-34–22 induces a human leukemic cell-specific cytotoxicity. Biochim Biophys Acta: Protein Struct Mol Enzymol 2003; 1547:57–63.

59. Cheong H, Gill SS. Cloning and characterization of a cytolytic and mosquitocidal delta-endotoxin from *Bacillus thuringiensis* subsp. *jegathesan*. Appl Environ Microbiol 1997; 63:3254–3260.

60. Thiery I, Delecluse A, Tamayo MC, Orduz S. Identification of a gene for Cyt1A-like hemolysin gene from *Bacillus thuringiensis* subsp. *medellin* and expression in a crystal-negative *B. thuringiensis* strain. Appl Environ Microbiol 1997; 63:468–473.

61. Mizuki E, Park YS, Saitoh H, Yamashita S, Akao T, Higuchi K, Ohba M. Parasporin, a human leukemic cell-recognizing parasporal protein of *Bacillus thuringiensis*. Clin Diag Lab Immunol 2000; 7:625–634.

62. Wasano N, Ohgushi A, Ohba M. Mannose specific lectin activity of parasporal proteins from a lepidoptera-specific *Bacillus thuringiensis* strain. Curr Microbiol 2003; 46:43–46.

63. Levinson B, Kasyan KJ, Chiu SS, Currier TC, Gonzalez JM. Identification of β-exotoxin production, plasmids encoding β-exotoxin, and a new exotoxin in *Bacillus thuringiensis* by using high-performance liquid chromatography. J Bacteriol 1990; 172:3172–3179.

64. Perani M, Bishop AH, Vaid A. Prevalence of β-exotoxin, diarrhoeal toxin and specific δ-endotoxin in natural isolates of *Bacillus thuringiensis*. FEMS Microbiol Lett 1998; 160:55–60.

65. Cherif A, Chehimi S, Limem F, et al. Detection and characterization of the novel bacteriocin entomocin 9, and safety evaluation of its producer *Bacillus thuringiensis* ssp. *entomocidus* HD9. J Appl Microbiol 2003; 95:990–1000.

66. Phelps RJ, McKillip JL. Enterotoxin production in natural isolates of Bacillaceae outside the *Bacillus cereus* group. Appl Environ Microbiol 2002; 68: 3147–3151.

67. Lee MK, Walters FS, Hart H, Palekar N, Chen JS. The mode of action of *Bacillus thuringiensis* vegetative insecticidal protein Vip3A differs from that of Cry1Ab δ-endotoxin. Appl Env Microbiol 2003; 69:4648–4657.

68. Siegel JP, Shadduck JA. Safety of microbial insecticides to vertebrates-humans. In: Laird M, Lacey LA, Davidson EW, eds. Safety of Microbial Insecticides. Boca Raton, FL, USA: CRC Press, 1990:101–115 (Chapter 8).

69. Federal Register. Viable spores of the microorgarnism *Bacillus thuringiensis* Berliner; exemption from the requirement of a tolerance, Title 40, Code of Federal Regulations (CFR), Part 180.1011, Federal Register 45:56347, August 25, 1980.

70. EPA. Health Effects Test Guidelines OPPTS 870.1100 Acute Toxicity Testing, 870 Series Testing Guidelines, U.S. Environmental Protection Agency, Washington, D.C., 2004 (http://www.epa.gov/opptsfrs/OPPTS_Harmonized/870_Health_Effects_Test_Guidelines/Drafts/870–1100.pdf).

71. EPA. Issues Pertaining to Bt Plant Pesticides Risk and Benefits Assessment, Presentation to the FIFRA Scientific Advisory Panel on Environmental Effects

and Final Report, October 18–20, 2000; accessed February 5, 2004 (http://www.epa.gov/oscpmont/sap/2000/october/brad3_enviroassessment.pdf and http://www.epa.gov/oscpmont/sap/2000/october/octoberfinal.pdf).

72. EPA. Characterization and Non-Target Organism Data Requirements for Protein Plant-Pesticides, Presentation to the FIFRA Scientific Advisory Panel by John Kough and Zigfridas Vaituzis and Final Report, December 9, 1999; accessed February 5, 2004 (http://www.epa.gov/scipoly/sap/1999/december/backgrd1.pdf and http://www.epa.gov/oscpmont/sap/1999/december/report.pdf).

73. http://www.isaaa.org/kc/CBTNews/press_release/briefs32/figures/global_area.jpg; accessed November 26, 2005.

74. OECD. Genetically Modified Foods: Widening the Debate on Health and Safety, the OECD Edinburgh Conference on the Scientific and Health Aspects of Genetically Modified Foods, OECD, Paris, 2000.

75. EPA. Microbial pest control agent data requirements. Title 40, Code of Federal Regulations (CFR), 2004, Table, Part 158.740.

76. Dudek BR, Hammond BG, Nemeth MA, Lemen JK, Astwood JD. 13-week feeding study in rats fed grain from YieldGard Event MON 810 corn. Toxicologist 2002; 66(1S) Abstract No. 930.

77. Mendelsohn M, Kough J, Vaituzis Z, Matthews K. Are *Bt* crops safe? Nat Biotechnol 2003; 21:1003–1009.

78. Noteborn HPJM, Rienenmann-Ploum ME, van den Berg JHJ, Alink GM, Zolla L, Kuiper HA. Food safety of transgenic tomatoes expressing the insecticidal crystal protein Cry1Ab from *Bacillus thuringiensis* and the marker enzyme APH(3)II. Med. Fac. Landbouww. Univ. Gent. 58/4b., 1993.

79. Noteborn HPJM, Rienenmann-Ploum ME, van den Berg JHJ, et al. Safety assessment of the *Bacillus thuringiensis* insecticidal crystal protein Cry1A(b) expressed in transgenic tomatoes. In: Engel KH, Takeoka GR, Teranishi R, eds. American Chemical Society Series 605. Washington, D.C., 1995:134–147.

80. Meher SM, Bodhankar SL, Arunkumar, Dhuley JN, Khodape DJ, Naik SR. Toxicity studies of microbial insecticide *Bacillus thuringiensis* var. *kenyae* in rats, rabbits and fish. Int J Toxicol 2002; 21:99–105.

81. Betz FS, Hammond BG, Fuchs RL. Safety and advantages of *Bacillus thuringiensis*-protected plants to control insect pests. Reg Toxicol Pharmacol 2000; 32:156–173.

82. Vazquez-Padron RI, Moreno-Fierros L, Neri-Bazan L, et al. Characterization of mucosal and systemic immune response induced by Cry 1AC protein from *Bacillus thuringiensis* HD 73 in mice. Braz J Med Biol Res 2000; 33:147–155.

83. Berberich SA, Ream JE, Jackson TL, et al. Safety assessment of insect-protected cotton: The composition of the cottonseed is equivalent to conventional cottonseed. J Agric Food Chem 1996; 41:365–371.

84. Redenbaugh K, Hiatt W, Martineau B, Emlay D. Determination of the safety of genetically engineered crops. In: Genetically Modified Foods. Washington, D.C, USA: ACS Press, 1995.

85. NIEHS Conference Proceedings. Genetically Modified Foods (Mini-Monograph). Environ Health Perspec 2003; 111:1110–1141 (http://ehp.niehs.nih.gov/docs/2003/111-8/toc.html).

86. Blanco C, Carrillo T, Castillo R, Quiralte J, Cuevas M. Latex allergy: clinical features and cross-reactivity with fruits. Ann Allergy 1994; 73:277–281.
87. Metcalfe DD. Introduction: What Are the Issues in Addressing the Allergenic Potential of Genetically Modified Foods? http://ehp.niehs.nih.gov/docs/2003/5810/abstract.html.
88. Bernstein JA, Bernstein IL, Bucchini L, et al. Clinical and laboratory investigation of allergy to genetically modified foods (http://ehp.niehs.nih.gov/docs/2003/5811/abstract.html).
89. Bannon G, Fu T-J, Kimber I, Hinton DM. Protein digestibility and relevance to allergenicity (http://ehp.niehs.nih.gov/docs/2003/5812/abstract.html).
90. Herman RA, Schafer BW, Korjagin VA, Ernest AD. Rapid digestion of Cry34Ab1 and Cry35Ab1 in simulated gastric fluid. J Agric Food Chem 2003; 51:6823–6827.
91. Fu TJ, Abbot UR, Hatzos C. Digestibility of food allergens and non-allergenic proteins in simulated gastric fluid and simulated intestinal fluid—a comparative study. J Agric Food Chem 2002; 50:7154–7160.
92. Hileman RE, Silvanovich A, Goodman RE, et al. Bioinformatic methods for allergenicity assessment using a comprehensive allergen database. Int Arch Allergy Immunol 2002; 128:280–291.
93. Kleter GA, Peijnenburg AACM. Screening of transgenic proteins expressed in transgenic food crops for the presence of short amino acid sequences identical to potential IgE-linear epitopes of allergens. BMC Struct Biol 2002; 2:8 (http://www.biomedcentral.com/1472–6807/2/8).
94. Kimber I, Dearman RJ, Penninks AH, Knippels LMJ, Buchanan RB, Hammerberg B, Jackson HA, Helm RM. Assessment of protein allergenicity on the basis of immune reactivity: animal models (http://ehp.niehs.nih.gov/docs/2003/5813/abstract.html).
95. Germolec DR, Kimber I, Goldman L, Selgrade MJ. Key issues for the assessment of the allergenic potential of genetically modified foods: Breakout Group Reports (http://ehp.niehs.nih.gov/docs/2003/5814/abstract.html).
96. Selgrade MJK, Kimber I, Goldman L, Germolec DR. Assessment of allergenic potential of genetically modified foods: an agenda for future research (http://ehp.niehs.nih.gov/docs/2003/5815/abstract.html).

10

Alternatives in Toxicology

Paul A. Locke and Alan M. Goldberg

Department of Environmental Health Sciences, Bloomberg School of Public Health, The Johns Hopkins University, Baltimore, Maryland, U.S.A.

INTRODUCTION

Alternatives in toxicology are broadly defined as toxicological and/or pharmacological tests or experimental approaches that alone or in combination (*i*) do not utilize animals, or replace the use of higher species animals with lower species animals to obtain equivalent or better scientific information, (*ii*) utilize fewer animals to produce equivalent or better scientific results than did previous animal-intensive testing, and (*iii*) minimize pain and distress to experimental animals. Alternatives have a distinguished history within toxicology (1). An informal survey of papers and posters presented at scientific meetings indicates that substantial scientific resources are devoted to the development and use of alternatives (2).

This chapter examines the vital and growing field of alternatives by illustrating the developments in chemical toxicity testing for regulatory purposes. Regulatory toxicology seeks to resolve societal concerns about compounds that are used, or are proposed to be used, in commerce, medicine, and other settings (3). Heightened concern over toxic releases is increasing the need for better data about thousands of compounds. At the same time, new toxicological techniques—alternatives—may be able to pinpoint where chemicals interact at the cellular or genetic level, and accelerate the process of identifying adverse effects (4). Without the development and implementation of alternatives, a huge increase in the use of animals would be necessary to answer important societal questions. Alternatives can help provide scientific answers to important toxicological questions without escalating animal use.

THE GROWING SOCIETAL NEED FOR INFORMATION ABOUT TOXIC CHEMICALS

Even before the Earth Day—dating back at least to the publication of *Silent Spring* over 40 years ago—people in the United States have been concerned about the release of toxic chemicals into the environment and the public health risks they create (5). Regulation of toxic substances became a part of environmental law in 1976, when the Toxic Substances Control Act (TSCA) was passed. TSCA was intended to protect human health and environment against the risks created by the introduction into commerce of new chemical substances (6). However, TSCA has not proved effective in generating the type and scope of information that society needs to evaluate risks about compounds already in commerce. TSCA does require that manufacturers submit to the Environmental Protection Agency (EPA) all toxicity data available, but it does not require that safety testing be undertaken. It is very difficult for EPA to use TSCA to evaluate, control, or, if needed, eliminate the thousands of compounds already in commerce (6).

Today, almost three decades after the passage of TSCA, our knowledge about toxic substances in the environment is still deficient (7). We know less than we should about the extent of human and environmental exposure to chemicals, and little about their toxic effects. Laws such as the Emergency Planning and Community Right-to-Know Act, passed in 1986, have helped fill some of the gaps in our understanding regarding exposure and potential exposure to compounds (8). But our knowledge about toxic effects has advanced only haltingly.

Since the 1980s, reports and investigations by several groups, including the EPA, the National Academy of Sciences, National Research Council, the Office of Technology Assessment, the Environmental Defense, and the American Chemistry Council, have demonstrated that a vast majority of chemicals produced and released in the environment lack basic toxicological data (7,9–12). Absence of this information, in turn, means that we cannot adequately assess whether these compounds create unacceptable environmental and public health risks (7).

But in the last five years, there has been a renewed societal effort to obtain more information about the effects of exposure. The chemical industry has agreed to work with the EPA and the environmental community to collect information about the toxicity of the approximately 3000 compounds produced in high volume—greater than one million pounds per year. The high production volume (HPV) program has obtained commitments from over 400 companies to test over 2000 compounds (13). A new Voluntary

Children's Chemical Evaluation Program has been launched around the same time, which will obtain health effects and exposure information on 23 pilot chemicals to which children are believed to be disproportionately exposed. EPA will use its authority under TSCA to demand testing if no companies agree to test voluntarily (14).

In Europe, the European Union is considering its own variant of a high volume testing and registration program called Registration, Evaluation, and Authorization of Chemicals (REACH). This program targets the same lack of knowledge about toxicity that the HPV and Children's Chemical Evaluation Program aim to correct. REACH would establish a system for assessing both existing and new chemicals. It contains three elements. First, basic information for approximately 30,000 substances (new and existing substances being produced in excess of 1 t/yr) would require submission of basic toxicology information into a common database. Second, for compounds produced in excess of 100 t/yr or certain other compounds of concern (about 5000 substances), a more extensive evaluation would be undertaken, looking particularly at long-term exposure. Third, compounds that are carcinogenic, mutagenic, or toxic to reproduction, and persistent compounds, would require authorization for use (15).

Extensive toxicology testing could eliminate the absence of basic toxicology information about the compounds that are of concern to these programs. There is general agreement internationally about the type and nature of data needed for identification of hazard. The Organization for Economic Cooperation and Development (OECD) has established six basic endpoints that should be studied to assess a chemical's toxicity for purposes of the HPV program. These endpoints are acute toxicity, chronic toxicity, developmental and reproductive toxicity, mutagenicity, ecotoxicity, and environmental fate. This is known as the screening information data set (16).

THE EVOLUTION OF ALTERNATIVES IN TOXICOLOGY

Combating toxic ignorance is a difficult challenge that will require the generation of substantial toxicologic information. It comes at a time when the practice and science of toxicology are evolving. Understanding of biological processes has advanced. While learning more about how chemicals and other stressors affect the environment and humans, toxicologists have uncovered relationships between exposure and disease that are rarely straightforward and can be perplexing. This has led to an effort at the EPA and other federal agencies, such as the National Institutes of Health, to examine the cellular, molecular, and gene-level changes caused by chemical exposures that lead to disease (17). Toxicology has moved from looking at overt disease endpoints—such as evidence of tumors in animals, observations about behavioral changes, or death—to the study of the actual processes that can cause such endpoints. In other words, toxicology has begun to focus its efforts on more subtle, precursor events that signal changes that lead to disease and occur before overt disease is present (18).

For preventing disease, and for designing more effective treatments, an understanding of the continuum from a healthy state to a diseased one is crucial. If science can explain the steps in this progression, practitioners can intervene earlier along this continuum. The earlier the intervention, the better the chance of preserving or restoring health. This change toward studying the "mechanism" or "mode" of action of compounds has been reflected in the latest risk assessment guidelines used by EPA and other federal agencies (19).

Where toxicologists were once comfortable using information about death or obvious signs of toxicity in regulatory decision-making agencies, they now face complex inquiries about the interaction among environmental contamination, health, and other stressors. At the same time, the toxicology toolbox has gotten much more sophisticated. Our national effort to decode the human genome and genetic codes of other species is beginning to bear fruit, and is being applied to environmental toxicology. The National Institute of Environmental Health Sciences (NIEHS) has initiated several ambitious projects that highlight the potential value of these tools. In 1998, NIEHS started an Environmental Genome Project (EGP) (20). The mission of the EGP is to improve understanding of human genetic susceptibility to environmental exposures. This project has focused its efforts on the molecular level to look at how individuals with genetic polymorphisms could be differentially susceptible to disease. In 2000, NIEHS created the National Center for Toxicogenomics (21). Toxicogenomics is a combination of traditional toxicology with genomics—the investigation of genetic make up and how it translates into biological functions. Toxicogenomics also marries data-rich genomics with toxicological endpoint evaluation and computational approaches (4). The Center is coordinating nationwide research to develop a toxicogenomics knowledge base.

Toxicogenomics uses microarrays, which are tools that can sift through and analyze information contained in a genome. A microarray consists of different nucleic acid probes that are chemically attached to a substrate (i.e., a microchip, glass slide, or microsphere-sized bead). By washing a toxic substance in solution over the microarray, any section of DNA affected can be made to glow or fluoresce, indicating the genes that have been altered.[a] Further investigation about whether this alteration causes adverse effects in living tissues would be warranted (22).

Toxicology's expanding ability to reach into the inner workings of molecular machinery has frequently outpaced the understanding of the

---

[a] For a more detailed discussion of microarrays, please see Chapter 7 of this work.

information obtained. Nevertheless, these new techniques are exciting because of the potential information they can yield. Advanced scientific techniques have the potential to illustrate the modes of action, and mechanisms by which the compounds in the environment cause or contribute to disease. At present, few, if any, undergo regulatory toxicology testing. Today, most basic tests rely on toxicology techniques that have not been updated for many years. This is especially true of tests that have been accepted by the EPA and OECD, and are used to make regulatory judgments (23).

HUMANE SCIENCE AND ANIMAL WELFARE

As the new multinational movement to expand testing of chemicals for toxicity gains speed, and promising advances in toxicologic methods develop, there are growing societal concerns about how animals are used in toxicology testing, including recognition of the ethical issues associated with the use of animals in science.

In Europe, animal welfare issues have been routinely addressed in testing programs and proposals. Excluding transgenic animals, the number of research animals used in countries like the United Kingdom has halved since the 1970s, in spite of the increase in products (24). This is also true in the United States (25). Animal use has dropped in the United States and the world since 1976. One of the seven objectives of the REACH program is to promote the use of nonanimal testing (15). But toxicology tests, especially those set out in U.S. testing protocols for basic toxicity information, require high numbers of animals, especially rodents such as mice and rats.

Research animals can feel pain and suffer distress. Ethicists have long studied the circumstances under which animals are used for cosmetic, pharmaceutical, and biomedical testing. Some have questioned whether it is acceptable, from a utilitarian perspective—the greater good argument (26). Other philosophers have argued that animals should be eliminated as biomedical research subjects, further asserting that animals should have rights that deserve protection under the law (27).

If today's approved regulatory testing protocols, were applied, widespread testing for chemical toxicity would require large numbers of animals (28). A basic battery of tests on one pesticide can involve more than 8000 animals across multiple species (Dow Chemical Company personal communication, 2002). The United Kingdom's Medical Research Council estimated that to fully implement the REACH program might involve the testing of as many as 30,000 compounds. This testing could cost approximately $11 billion and sacrifice 13 million animals (Thomas Hartung, ECVAM, personal communication). EPA estimated in 1998 that the full battery of basic toxicology tests could cost up to $205,000 per compound; and such testing could reveal the need for more detailed testing.

Today, at least for biomedical research, about 70% of the population supports the use of animals for necessary studies. This high level of support is understandable based on the belief that animal experimentation is necessary for eliminating human disease and suffering. It is likely that this rationale carries over to animal testing associated with chemicals in the environment, because the ultimate aim of this research is to prevent disease and protect health. But the public is increasingly expressing concern about the numbers of animals used, seeking nonanimal testing and demanding that when animals are used, they are treated humanely and that their pain and distress be eliminated or at least minimized.

The 3Rs

Understanding animal welfare issues—coupled with an appreciation for the challenges and opportunities in regulatory toxicology—is the starting point for gaining an appreciation of alternatives in regulatory toxicology. Although ethical and economic arguments are important, applying good animal welfare to toxicology research—practicing humane science—will help produce high-quality, reliable data for regulatory decision making. It is also important in bridging the gap between traditional, animal-intensive toxicology and the new technologies that toxicology is rapidly developing. Bringing these new toxicological tests online has the potential to jointly reduce animal use and produce information that can aid in regulatory decision making. In addition, animal-based toxicology research is improved when principles of humane science are applied. To explain why humane science yields robust and reliable results, a fuller understanding of what "humaneness" is and how it applies in the laboratory is needed. In short, toxicologists should come to recognize, appreciate, and implement a series of concepts known as the "3Rs."

In 1959, two British scientists, William Russell and Rex Burch, published *The Principles of Humane Experimental Technique*, a book that defines and explains humane science (29). Humane science is captured in the concepts of refinement, reduction, and replacement, which are referred to as the 3Rs. Alternatives to animal tests are a key part of the 3Rs framework. The 3Rs are based equally on ethical considerations of animals in the laboratory and the recognition that when the researcher in experimental design and implementation appropriately applies these principles, they result in experimental protocols that are likely to produce more robust scientific results.

The first of the 3Rs, refinement, is defined as any method that reduces or eliminates pain and distress in animals during experiments (1). To implement the refinement prong of the 3Rs, it is not enough to simply administer analgesics or anesthesia to animals in pain. Every procedure in the experimental protocol must be considered from the perspective of the need to reduce or eliminate pain and distress. Thus, noninvasive imaging—the use of magnetic resonance imaging (MRI), positron emission tomography (PET) scanning, X-ray techniques, or biophotonic imaging—is finding its way into the laboratory (30). Biophotonic imaging is possible when luciferase, a compound obtained from fireflies or glowworms, is attached to cells, bacteria, viruses, or specific genes. When these genes are "turned on," indicating a molecular change of toxicological significance, a luminescence reaction takes place. Because light is produced, the reaction can be measured without invasive procedures. Toxicogenomics and related techniques fall squarely in this area.

Refinement also encompasses the substitution or use of species lower on the phylogenic scale for species that are higher in phylogeny. For example, using rodents instead of primates in an experimental protocol to obtain the same or more scientific information is an example of refinement, as is the use of *Caenorhabditis elegans* or zebra fish instead of rodents. Finally, designing experiments that study humane endpoints, such as early tumor growth or changes in biological or genetic markers, rather than death, embraces refinement because the animals that are studied are likely to be healthier, and in less pain and distress, during early stages of disease.

The second of the 3Rs, reduction, is defined as a method that seeks to use fewer animals in an experimental protocol to obtain the same or similar information of scientific value, or use the same number of animals to obtain more scientifically valuable information (1). For example, using a methodology such as biophotonic imaging can reduce the numbers of animals used because each animal can serve as its own control. In addition, reduction can be achieved when animal tissues or samples are used in more than one scientific protocol.

Russell and Burch point out that "one general way in which great reduction may occur is by the right choice of strategies in planning and performance of whole lines of research." Thus, statistics, computational methods, and design considerations are critical tools in an overall approach to humane science (31).

Replacement, the third of the 3Rs, is defined as the use of techniques that do not use living animals (1). Sophisticated methods being developed at NIEHS substitute in vitro protocols for animal testing. Structure–activity relationships and computational methods use information about the molecular structure of a compound to determine its potential for harm. The rapid advances in genomics and proteomics hold great promise as replacement strategies. Databases containing genomics information are yet to be widely utilized, but their value to toxicology in general and alternatives in particular are not in doubt. In addition, the ability to study a very specific genetic change that is known to be, or suspected of being, important in the development of disease, or a change that indicates that a movement toward disease has, or is beginning, to take place (as can be examined in microarrays or DNA chips), can be powerful in showing the early effects of exposure to compounds.

Increasing Scientific Robustness by Practicing Humane Science

By focusing on humane scientific techniques, Russell and Burch also made explicit what biomedical researchers had long recognized in practice: that good animal welfare sets the stage for better scientific results. Since publication of their book, there is a growing recognition that high standards of animal welfare result in scientific research that is more reliable, relevant, and reproducible—better science, in other words (30).

Consider, for example, the role that stress can play in animal-based toxicology. Physiologically, vertebrate animals under stress exhibit elevated heart rates, higher blood pressure, and increased body temperature. Crudely put, their physiological engines are revved up because of the stress they are under. When data are obtained from animals under stress, it is fair to ask whether such data are truly representative of the responses to exposure that they would experience in settings that more closely parallel typical environmental exposures. If the answer to this inquiry is no, it is harder to argue that this information can be extrapolated to the exposures that are of interest to a regulator (32).

Another example of this principle is spelled out in some of the experimental work in animal behavior, including research into stereotypy. Stereotypy, defined as repetitive, functionless, nonvarying actions, is fairly common, especially in laboratory animals raised in a nonenriched laboratory environment (33). Behavioral research in this area is growing rapidly, and seems to indicate that stereotypy can result in physiological changes, especially in neurological chemistry and pathways. Such changes are hypothesized to have an unknown but potentially relevant effect on data generated using such animals (33).

It is not yet possible to determine with certainty whether these changes could affect underlying experimental results and conclusions. Nevertheless, given the possible risks associated with incorrect conclusions due to suboptimal data, researchers should see the benefit in eliminating, to the extent possible, the conditions that lead to excess or unnecessary stress or stereotypic behavior. As toxicologic phenomenon under study become more complex and mechanistic, practicing humane science becomes even more important.

Institutional Mechanisms—Federal Humane Science Law and Regulations

In the United States, there is an evolving body of humane science law, including the Animal Welfare Act (AWA), which defines the relationship between the laboratory researcher and certain, but not all, experimental animals (AWA, 7 USC 552131–59). The AWA requires that certain research animals are handled in a humane way and are provided with appropriate, housing and feeding. It contains provisions calling for the minimization of pain and distress in experimental procedures, the review of alternatives to any procedure that can produce pain and distress to animals, and the rendering of veterinary care to animals so that they have the benefit of tranquilizers, analgesics, and anesthetics when necessary [AWA §2143(b)].

In addition, the AWA requires the establishment of an Institutional Animal Care and Use Committee (IACUC) to monitor animal research and self-police the provisions of the AWA at facilities covered by the Act [AWA §2143(b)]. In addition to their other activities, IACUCs must appoint a minimum of three members; one must be a veterinarian and one must be a person who is not affiliated with the institution and who represents community interests. IACUCs are also required to inspect, at least semiannually, all animal study areas and animal facilities with a specific eye toward practices involving pain and distress, and provide training for scientists in humane practice and procedure. Perhaps most importantly, IACUCs must review and approve research protocols involving animals to assess compliance with the AWA.

Rats, mice, and birds bred for use in research—which make up close to 90% of all animals used in biomedical research—fall outside the requirements of the AWA [AWA §2132 (g)]. However, the United States Public Health Service (PHS) (of which the National Institutes of Health are a part) has adopted an animal protection policy that has much in common with the AWA, but with a broader reach (34). The PHS policy covers all institutions that receive PHS support and applies to all vertebrate animals. Between the AWA and the PHS policy, it is likely that the majority of research facilities that use vertebrates in research are covered. It is not possible to determine with precision the percent or number of facilities covered because such statistics are not available for the United States.

ASSESSING ALTERNATIVES AND HUMANE SCIENCE: SUCCESSES, CHALLENGES, AND THE FUTURE

Society's need for more toxicological information and the growing sophistication of toxicological science complement and embrace the practice of humane science. To fully implement a humane science approach to regulatory toxicology, a proactive science policy should be implemented that takes advantage of advances in toxicology and is based on the highest of ethical standards. The 3Rs and a humane science approach to research chart the path for accomplishing these important societal goals.

Successes

Sometimes like beauty, success is in the eye of the beholder. However, there is hard evidence of major successes in the field. The Interagency Coordinating Committee on the Validation of Alternative Methods (ICCVAM) has validated two methods—the local lymph node assay and corrosivity testing (35). The local lymph node assay has been generally and widely accepted throughout the scientific community and is used on a regular basis. There has been establishment of alternatives infrastructure including, but not limited to, the creation of ICCVAM, a coordinating committee of 14 different government agencies, and the establishment of the Animal Welfare Information Center at the USDA, serving the scientific community at large and providing references, literature searches, and general access to available knowledge. Most importantly, animal testing has essentially been eliminated in the testing of household goods, consumer products, and in the personal care industry. This was accomplished over a 20-year period and is clearly recognized as one of the most visible series of accomplishments in the field. However, there are many more accomplishments that provide for additional long-term stability and growth in this area. The establishment of the World Congress on Alternatives and Animal Use in the Life Sciences will shortly hold its fifth triennial meeting. The anticipated audience will come close to 1000 people and represent countries all over the world. In Europe, the authors believe, it is possible to witness more than one meeting and probably closer to two or three meetings a month dealing with the issue (this is not necessarily the case in the United States). One of the more significant events is the teaching of humane science courses, such as the one offered online and prepared by faculty at the Johns Hopkins University associated with the Center for Alternatives to Animal Testing, Comparative Medicine, The Animal Care & Use Committee and the Provost's office (http://coat.jhsph.edu/humanescience/login.cfm). More than 300 students have taken the course during the first year. As the material in this course and others get disseminated, the practice of humane science will become as routine as the use of a pH meter in a laboratory.

In spite of these successes, it is not likely that nonanimal tests will completely replace animal tests in regulatory toxicology in the near future. When replacement alternatives either cannot be developed or do not provide needed information, or when nonanimal tests will not suffice for other reasons, animal testing will occur. In such cases, the animal tests should be carried out in as humane a way as possible, recognizing that the humane approach fosters good scientific data.

Challenges

Bringing better science into the regulatory process by developing and using more alternatives and practicing humane science faces certain challenges. Validation of alternatives is the first major hurdle. There exists today a series of animal-based toxicology tests that are traditionally used to make regulatory decisions (36). Some of these tests are codified in regulations, such as EPA test rules under TSCA or OECD testing protocols, or both. If the intent of alternatives is to replace these tests, alternatives must demonstrate that they produce predictive and repeatable results. In regulatory toxicology, validation is defined as the establishment of the reliability and relevance of an alternative method for a specific purpose. In other words, for a test to be validated, it must show that it can consistently reproduce results and that these results are predictive—that is, they provide information that is of value for assessing whether the compound is a hazard (37).

Validation

A validation process from test development to acceptance is set out in Figure 1. As discussed above, a validation process for alternatives has been established in the United States by the ICCVAM, and several international conferences have been devoted to validation (37–39). This process is complex, expensive, and time consuming. For validation, among other things it must be shown that the endpoint of the alternative is related to the biological effect of interest (e.g., showing biological relevance to the toxic process in question), the result is reproducible among different laboratories, and the methodology has been published in (or prepared for publication in) a peer-reviewed journal.

Bringing an alternative through the validation process is a daunting and expensive task. Based on experiences in validating tests in Europe, validation studies can cost as much as $1.6 million and take approximately 10 years from test development to final validation. As discussed previously, approximately two alternative tests have been validated by ICCVAM and 10 by European Centre for the Validation of Alternative Methods (ECVAM), its European counterpart (40). Clearly, this process needs to be streamlined so that it is faster and less costly, while delivering the same quality of information.

Regulatory Acceptance

After a test has been validated, it must then jump a second major hurdle. Regulatory agencies must formally accept the test for use in their programs. ICCVAM guidance on regulatory acceptance states that such acceptance is predicated on several factors, including whether the method has undergone independent scientific peer review, contains a detailed protocol with standard operating procedures, is robust in methodology and transferable among laboratories, and generates information that assists in risk assessment. Some of these criteria overlap with the ICCVAM criteria for validation.

In the United States, validation does not guarantee regulatory acceptance. The ICCVAM law does not require that agencies adopt validated alternatives; each agency can decide for itself whether a validated alternative test will be acceptable under its regulatory programs (41). In addition, agencies are under no obligation to replace the tests already used with the validated alternatives (42 USC §2851-4). In fact, it is possible that agencies could require both tests.

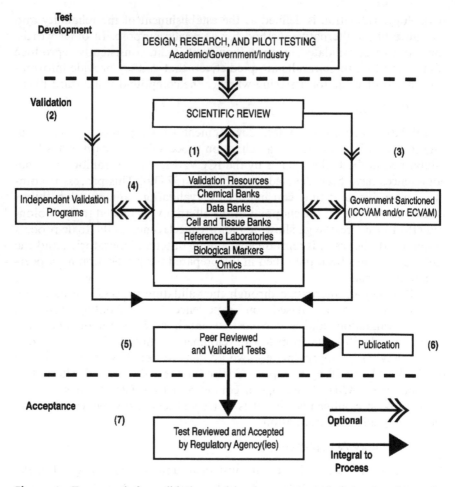

Figure 1   Framework for validation and implementation of alternatives in regulatory toxicology.

International Harmonization

A third possible hurdle exists. In addition to having alternatives that are validated and accepted by the U.S. regulatory agencies, international organizations, specifically ECVAM and OECD, must accept these tests if they are to have maximum impact in advancing toxicologic testing, regulating dangerous compounds, reducing animal use, replacing animals, and minimizing pain and distress. In some cases, international acceptance proceeds contemporaneously with acceptance in the United States. International and U.S. agencies are also working to adopt and use the same protocols for testing, thereby harmonizing the toxicological requirements for alternatives. For example, EPA's Pesticides Office has been working with OECD on common issues.

CONCLUSIONS

The history of regulatory toxicology and its recent developments strongly suggest that a humane science approach and the development, validation, and use of alternatives (especially replacement alternatives, when feasible) hold great promise for advancing our toxicological knowledge base about chemicals in the environment. To assure that progress continues to be made in regulatory toxicology science and against toxic substances exposures, we must utilize techniques and adopt a regulatory agenda that fosters and encourages replacement alternatives.

A focus on replacement alternatives to animal tests, and the full implementation of a humane science agenda, will serve us well, both as a nation and as a participant in an international community concerned about the health hazards associated with chemicals in the environment and the ethical treatment of animals in toxicological research.

REFERENCES

1. Russell WMS, Burch RL. The Principles of Humane Experimental Technique, 1959 (reprinted as a special edition, 1992) available at http://altweb.jhsph.edu/publications/humane-exp/forward.htm
2. Han J, Hooser G. Handbook of Laboratory Annual Science. Vol. 1. Boca Raton: CRC Press, 2003.
3. Marquardt H, Schäfer S, McClellan R, Welsch F. Toxicology. Elsevier 1996.
4. Frueh FW, Huang SM, Lesko L. Regulatory Acceptance of Toxicogenomics Data (Editorial). Environmental Health Perspectives 2004; 112(29):A663.
5. Carson R. Silent Spring, 1962.
6. Brown E, Hatcher J. A Practitioner's Guide to the Toxic Substances Control Act, TSCA Deskbook, Environmental Law Institute, 1999.
7. Environmental Defense. Toxic Ignorance: The Continuing Absence of Basic Health Testing for Top Selling Chemicals in the United States, 1997.
8. Burcat JR, Hoffman AK. The Emergency Planning and Community Right-to-Know Act of 1986: An Explanation of Title III of SARA. Environmental Law Reporter 1988; 18:10007–10027.
9. EPA. Chemical Hazard Data Availability Study: What Do We Really Know About High Production Volume Chemicals? (Office of Pollution Prevention & Toxics, 1998) Available at www.epa.gov/oppt/chemtest/hazchem.htm; (last accessed December 16, 2004).

10. National Research Council. Toxicity Testing: Strategies to Determine Needs and Priorities, 1984.

11. US Congress, Office of Technology Assessment. Identifying and Regulating Carcinogens, OTA–BP–H-42 (Washington, D.C., November 1987) (Available at www.wws.princeton.edu/~ota).

12. www.americanchemistry.com/randt.nsf/unid/unar-4dfn3h

13. www.epa.gov//chemrtk

14. 65 Federal Register 81686 (December 26, 2000) [HPV program request for adoption].

15. Commission of the European Communities. White Paper: Strategy for a Future Chemicals Policy, February 27, 2001.

16. Green S, Goldberg A, Zurlo J. TestSmart–high production volume chemicals: an approach to implementing alternatives into regulatory toxicology. Toxicol Sci 2001; 63:6–14.

17. www.nih.niehs.gov/ntp

18. www.nihroadmap.nih.gov/

19. EPA. Risk Assessment Guidelines (Available at www.epa.gov/ncea).

20. www.niehs.nih.gov/envgenom/home.htm

21. www.niehs.nih.gov/net/home.htm

22. www.niehs.nih.gov/net/glossary.htm (last visited October 10, 2004).

23. www.altweb.jhsph.edu/regulations/EPA/public-drafts.htm (last visited December 17, 2004)

24. BBC News. Animal Tests See Steady Decline (http://ness.bbc.co.uk/go/pr/fr/-/1/hi/sci/tech/3650057.stm)

25. Gauthier C. Overview and analysis of animal use in North America. ATLA 2004; 32(suppl 1):275–285.

26. Singer P. Animal Liberation, 1975.

27. Regan T. The Case for Animal Rights, University of California Press, 1985.

28. DeGreeve P, DeLeeuw W, van Zutphen BFM. Trends in Animal Use and Animal Alternatives. ATLA 2004; 32(suppl 1):13–19.

29. www.altweb.jhsph.edu/publications/humane-exp/forward.htm

30. Stokstad E. Humane science finds sharper and kinder tools. Science 1999; 286:1068–1071.

31. Vaughan S. Optimising resources by reduction: the FRAME reduction committee. ATLA 2004; 32(suppl 1):245–248.

32. Coenraad FM, Hendriksen, Cavid B, Morton, eds. Humane endpoints in animal experiments for biomedical research. Proceedings of the International Conference, 22–25 November 1998, Zeist, The Netherlands, 1998.

33. Garner JP, Meehan CL, Mench JA. Stereotypies in caged parrots, schizophrenia and autism: evidence for a common mechanism. Behav Brain Res 2003; 145:125–134.

34. US Public Health Service, US Government Principles for the Utilization and Care of Vertebrate Animals Used in Testing, Research and Training (Available at http://grants.nih.gov/grants/olaw/references/phspol.htm#USGovPrinciples).

35. http://iccvam.niehs.nih.gov/methods/review.htm

36. PPTS Harmonized Test Guidelines, Series 810–885 (Available at http://epa.gov/OTTPS_Harmonized/).

37. Leon B, Gregory C, Rodger C, Mark C. Validation of Alternative Methods for Toxicity Testing. Environmental Health Perspective 1998; 106(S2).
38. Goldberg AM, Frazier JM, Brusick D, et al. Report of the Validation and Technology Transfer Committee of the Johns Hopkins Center for Alternatives to Animal Testing. Framework for validation and implementation of in vitro toxicity tests. Xenobiotica 1993; 23(5):563–572.
39. Manfred L, Horst S. Currently available in vitro methods used in the regulatory toxicology. Toxicol Lett 2002; 127:127–134.
40. http://ecvam.jrc.cec.eu.int/index.htm
41. Letter from Joseph Merenda, EPA, to Ken Olden, NIEHS, dated 12/02/2003, regarding the "up and down" procedure.

11

Genetic Toxicology and Regulatory Policy

David Brusick

Bumpas, Virginia, U.S.A.

INITIATION OF GENETIC TOXICOLOGY TESTING

During the 1960s, several highly visible environmental health issues converged, simultaneously creating greater awareness in the general public of their surroundings and heightening their concern over exposure to the ever-increasing number of man-made chemicals released into the environment. The breadth and intensity of this concern was exemplified by the establishment of an international "Earth Day" in 1970. Shortly thereafter, two other events, the "War on Cancer" initiated in 1971 and the discovery of human toxicity associated with the Love Canal chemical waste site in 1977, contributed to the growing interest in the effects of environmental chemicals on genes and chromosomes.

Because Love Canal and the War on Cancer directed the public's attention toward the identification of chemicals in the environment that might be dangerous, the early 1970s were characterized by a burst of activity involving basic and applied research in genetics and cancer. Broad public and scientific interest in applied genetics ultimately led to the creation of organizations with environmental missions such as the Environmental Mutagens Society (EMS). These new organizations quickly created a set of expectations regarding the responsibility of government agencies to protect society from environmental chemicals that could alter an organism's genetic integrity.

These organizations focused on building the case that industrial development had become a "good news: bad news" scenario. The "good news" was produced by the flow of new technology and products, improving the quality of modern life. New medicines and devices were introduced which extended the life expectancy of the general population. The "bad news" part of technological advancement was the potential to increase the incidence of chronic diseases following exposure to the by-products of industrialization. Scientists were finding that long-term, low-level exposures to environmental chemicals were associated with a number of serious adverse health effects including cancer, fetal malformations, and reproductive failure (1–3).

A second revelation stemming from these activities was the finding that very few of the chemicals found in our environment had ever been tested for carcinogenic, teratogenic, or reprotoxic activity. Several tragic incidents, including the finding that thalidomide induced severe birth defects in children of women taking the drug and that diethylstilbestrol induced uterine cancer, raised doubts as to whether the array of safety tests relied upon by regulatory agencies to detect human toxins were reliable or adequate (4).

During the period 1970 to 1975, a large number of the scientific literature dealing with cancer and reproductive toxicity established strong support for the prediction that changes in specific genes and/or chromosomes were somehow associated with the initiation of these effects and testing for genotoxicity should be considered (5). In 1974, several contract testing organizations were in the process of commercializing some of the more common tests for genotoxicity, and mutagenicity testing was becoming formalized into frameworks and tier systems (6,7). The characteristics of these early testing strategies would undergo transformation several times during the 1970s.

At the outset, genetic toxicology's intended function was to identify environmental mutagens and clastogens that produced transmissible mutations in human germ cells, capable of giving rise to new diseases in the offspring. Consequently, testing strategies driving the initial structure of genetic toxicology were also based on the perceived need to protect the human gene pool from mutagens produced by industrial development. Selling this concept to health regulators was more difficult than expected, primarily owing to the absence of any documented human germ cell mutagens other than radiation. Adding to these difficulties was the absence of good animal models that permitted easy detection and confirmation of transmissible gene mutations. The only accepted tests for gene mutation detection at that time were those developed in the fruit fly or specific strains of mice (8,9). The fruit fly was not embraced as a test system whose data could be reliably extrapolated to humans, and the mouse models required specialized testing resources and the production of several thousand F1 offspring to detect even relatively potent germ cell mutagens.

In 1973, a paper published by Ames et al. (10) became a watershed in the emergence of genetic toxicology as a unique discipline for determining safety of chemicals. This and similar publications during that period produced a highly visible impact on cancer testing techniques by driving the proliferation of tests designed to detect agents that specifically interact with deoxyribonucleic acid (DNA) to produce mutation or other types of nucleic acid lesions (genotoxins).

Genetic toxicology quickly changed its focus from germ cells to somatic cells and the detection of possible chemical carcinogens. By 1980, more than 100 different test systems, mostly in vitro, had been devised to detect genotoxic carcinogens (11). The range of endpoints used to classify chemicals as genotoxic proliferated as well and included not only mutation and clastogenicity, but also sister chromatid exchange, induction of DNA repair, differential DNA repair, mitotic crossing over, mitotic recombination, aneuploidy, micronuclei formation, DNA strand breakage, and morphological cell transformation. Animal cancer data were being compared to the results of genetic tests used singly and as batteries. Application of genetic toxicology in safety testing was encouraged, principally owing to the initial high correlation between genotoxicity and carcinogenicity and by the discoveries that specific genes (oncogenes and tumor suppressor genes) were involved in tumor development (12–15).

Finally, in 1975, under the auspices of the EMS, a paper was published in Science (16). This publication attempted to convey the extent to which environmental chemicals might endanger the human genome and the importance of developing better test methods for detecting mutagens and carcinogens in the environment. This document, along with similar publications by various environmental organizations, primarily influenced the expansion of genetic toxicology beyond the research laboratory and made a strong case for proposals that production of and/or exposure to mutagenic and clastogenic chemicals should be regulated. The initial response to these proposals by the industrial and pharmaceutical manufacturers was cautious skepticism.

ACTIVITIES THAT FACILITATED THE ENTRY OF GENETIC TOXICOLOGY INTO THE REGULATORY ARENA

The absence of unequivocal evidence confirming the presence of environmentally induced mutation in human germ cells or the existence of unique germ cell mutagens that would slip through traditional safety assessments, provided the basis for arguments claiming that it was too early to add genetic tests to regulatory safety assessments. Among regulators and toxicologists, many believed that without explicit human epidemiological evidence linking genetic diseases to damage of the genome, it would be impossible to build defendable policies around regulation of agents on the presumption of their mutagenicity. Even though scientists had demonstrated the induction of germ cell mutations in mice by radiation and chemical agents, the processes to extrapolate those findings to the detection of new mutations in human populations had not been given adequate attention (17–19).

Traditional regulatory toxicology policies had always been built around the belief that identification and assessment of adverse health effects in animal models must be amenable to direct extrapolation to the human experience. A change in that paradigm was unlikely to be accepted by regulatory policy makers. If genetic toxicology were to be formally added to safety assessments, more information about the mechanisms and kinetics of germ cell mutation induction in mice and humans would be required.

Following the creation of the EMS in 1969, activities aimed at developing a better understanding of mutagenesis and its impact on human health were undertaken by regulatory as well as advisory organizations. One such activity was a pair of symposia sponsored by the National Institute of Environmental Health Sciences (NIEHS) and U.S. Environmental Protection Agency (EPA) (20). These symposia covered the emergence of in vitro testing and the role it could play in developing toxicology assessments. The output from this and similar activities contributed to the standardization of genetic testing. Ultimately, data assessment initiatives sponsored by the government and advisory organizations proved to be essential in demonstrating the potential value of genetic toxicology data in regulatory decision making. Four specific examples are discussed below.

Environmental Mutagen Information Center

One of the first activities initiated by the EMS was to build an international data repository. With funding from EPA and NIEHS, Environmental Mutagen Information Center (EMIC) was created at the Oak Ridge National Laboratory in 1970. This data center collected and disseminated genetic toxicology data to the interested investigators and regulators. EMIC began to compile data on the numbers and types of chemicals tested for genotoxicity, test method development or validation, and summaries of approaches for quantitative genetic risk estimates. It shortly became the international source for information about genotoxicity. Without this organization, our understanding of genetic toxicity would still be marginal.

EPA GENE-TOX (PHASES I AND II)

Once a critical mass of published data had been collected by EMIC, the logical next step was to use the information to answer questions about the utility of short-term tests in defining genetic hazard and risk or the relationship between genotoxic and carcinogenic activities of chemicals. In 1979, the EPA GENE-TOX program was established to provide a comprehensive review of genetic toxicology literature. The purpose of phase I of this program was to compile and analyze the available data to be used in establishing standard genetic testing practices for the evaluation and possible regulation of toxic substances.

The steps involved were as follows:

1. Prepare comprehensive data reviews for all routine test methods.
2. Recommend appropriate, "standard" testing protocols.
3. Develop and publish evaluation criteria for the recommended tests.

During 1998, over 3000 records were assessed under the GENE-TOX program. The program provided a massive amount of information to regulators who were looking to construct reliable testing schemes.

In the second phase of the GENE-TOX program, separate task groups were formed to assess the compiled data for purposes of defining the development status of various tests, determining the reliability of the test to identify germ cell mutagens, as well as the possible application of selected short-term tests as screens for chemical carcinogens (21–23). Although the overall results in the 3000 records were decidedly biased toward positive responses and there were substantial data gaps for some tests, the GENE-TOX information indicated that useful information about compound safety could be generated from genetic toxicology evaluations (24).

In 1986, the EPA published its Mutagenicity Risk Assessment Guidelines (25,26). These guidelines were structured with substantial reliance on data from the GENE-TOX program as well as input from other organizations such as International Commission for Protection Against Environmental Mutagens and Carcinogens (ICPEMC), the International Agency for Research on Cancer (IARC), and NIEHS. In their risk assessment guidelines, EPA proposed that genetic test data be evaluated with a "weight-of-evidence approach" using test responses as well as toxicokinetics and pharmacology to arrive at risk decisions. The guidelines also established a chemical classification system based on data quality (i.e., sufficient, limited, etc.) for mutagenic risk similar to that created for carcinogens by the IARC and the National Toxicology Program (NTP).

The proposed risk assessment approach used both explicit and intrinsic toxic properties of chemicals to determine their hazard classification (e.g., mutagenic risk required the demonstration that a chemical was a somatic cell or microbial mutagen and that it could be detected within the blood/gonad barriers). The difficulty in using the EPA guidelines was that the data used to classify chemicals were, for the most part, indirect (i.e., not derived from germ cell mutations) and assessed the consistency of the evidence but did not attempt to differentiate mutagenic agents according to a level of risk. Without quantitation of either absolute or relative risk for chemicals with positive test results, this approach consistently leads to unresolvable differences involving risk versus benefit.

ICPEMC

In 1977, ICPEMC was established. Because of its international representation, ICPEMC served as a forum where consensus views might emerge on the controversial issues related to mutation and cancer. These views were expected to facilitate recommendations that might serve as the basis for guidelines and, ultimately, regulations. For more than 15 years, this organization, using a series of task groups, committees, and subcommittees, produced scientific reports, reviews, and white papers on all aspects of mutagenesis and carcinogenesis. In 1983, ICPEMC Committee 3 published an extensive report (27). The members of the committee had all served in the regulatory capacity for European and North American countries. The recommendations put forth in the paper regarding the role of mutagenicity in regulatory policies are as follows:

- Mutagenicity should be considered an important component of the toxicological profile of regulated chemicals.
- Mutagenicity testing should be conducted in a hierarchical manner that facilitates efficient use of available resources.
- The selection of a regulatory response/action should take into account the mode of action by which a chemical produces an adverse effect (i.e., chemicals that react directly with DNA to initiate neoplasia carry a greater risk per unit of exposure than do promoters or epigenetic carcinogens).
- Short-term test data, in combination with more conventional toxicological tests, might be acceptable for making decisions that no additional testing is needed (i.e., supplemental data leading to an "acceptance" decision).

ICPEMC committee members failed to achieve a consensus on two important decision points. The first was the rigor with which a regulatory agency should require or request short-term test data to be submitted, and the second, very much related to the first, was the rigor with which a regulatory agency should respond to positive test results (i.e., restrict vs. ban). The reluctance to reach consensus by the committee members was linked to the issue discussed earlier of the inability of science to demonstrate a cause–effect relationship between chemical exposure and new mutations in the human gene pool. Without adequate proof of this linkage, calling for mutation testing or banning products based on their results would certainly initiate complex litigation.

NTP

NTP initiated a project, in 1984, to evaluate four commonly used short-term genetic toxicology assays for their ability to predict tumorigenicity in all of the chemicals evaluated in the NTP bioassay program. The tests included in this evaluation were the Ames test, for chromosome aberrations and sister chromatid exchange in Chinese hamster ovary cells, and an assay for gene and chromosome mutation at the thymidine kinase gene in L5178Y mouse lymphoma cells (mouse lymphoma assay). The stimuli for this program included the following:

1. The initial high correlation reported between the Ames test results and the results from rodent cancer bioassays.
2. Results from the EPA GENE-TOX program indicating that mammalian cell assays often responded to carcinogens that failed to induce reverse mutation in the Ames strains.
3. The discovery that gene and/or chromosome mutation was involved in the activation of proto-oncogenes and tumor suppressor genes.

The purpose of this program was to establish a short-term test database that could be used to identify and validate a test, or group of tests, that could accurately predict rodent carcinogenesis, at least on a qualitative basis. However, by 1990, the data emerging from this program were suggesting that animal carcinogenesis was a complex, multifactorial process that could not be accurately predicted by any of the four tests evaluated, whether combined or used alone (28,29). Later in the program, selected in vivo tests such as the rodent micronucleus assay were added to the evaluation, but data from these tests failed to significantly improve the predictability of short-term tests, in general (30). The lack of good concordance between genotoxicity and cancer observed in the NTP project raised questions about the applicability of short-term genotoxicity tests as predictors of carcinogenicity and as useful components of regulatory assessments. Prior to a comprehensive analysis of the data from the NTP studies, many international regulatory bodies, including the USEPA, had already added genetic screens into their testing guidelines and were requesting short-term tests as components of assessments before product release or approval.

At the outset, the role of genetic testing in regulatory review and approval of chemicals, food additives, and drugs was unclear and fragmented. While some tests were specified in all batteries (e.g., Ames), there were significant differences among the recommended test batteries for evaluating nonpharmaceutical agents. A discussion of tests and testing strategies for genetic toxicology can be found in volume 21 of the journal Environmental and Molecular Mutagenesis (31,32). Worldwide regulatory guidelines for genotoxicity testing with primary attention given to nonpharmaceutical products were also summarized the journal.

At about the time of the 1993 publication, the International Conference on Harmonization (ICH) of the technical requirements for registration of pharmaceuticals for human use established a Genotoxicity Working Group. This group was organized in 1991 and consisted of regulatory and nonregulatory members from the United States, the European Community, and Japan. The output from the working group consisted of two guidance documents. They were designated ICH S2A "Specific Aspects of Regulatory Genotoxicity Tests for Pharmaceuticals" and ICH S2B "Genotoxicity: A Standard Battery for Genotoxicity Testing of Pharmaceuticals." Acceptance of the ICH test strategy across international boundaries was complete by 1997. The actual test procedures recommended in ICH S2B were heavily dependent on guideline protocols published by the Organization for Economic Cooperation and Development (33). Refinements and modifications in the designs and acceptance criteria were made to the guidelines through the past several years. By the end of the twentieth century, mutagenicity testing guidelines for safety testing were found in most international regulatory schemes. The pattern of testing was relatively consistent whether the product was going to be approved in the United States, Canada, the European Union, or Japan. One aspect of genetic toxicology testing strategies that was not included in the guidelines, however, was a set of rules (logic) on how to incorporate short-term test responses into regulatory decision making.

GENETIC TOXICOLOGY TECHNOLOGIES AND CONCEPTS THAT INFLUENCED CURRENT REGULATORY PROCESS AND STRATEGIES

The Gene as a Relevant Target for Toxic Agents

Before 1970, virtually all toxicology tests were focused on finding adverse effects that would affect individuals or their fetuses exposed directly to the agents under evaluation. Genetic toxicology brought a new dimension to regulatory toxicology by the end of the 1970s. The "golden age" of molecular biology, during the 1960s, provided a wealth of understanding about the structure, regulation, and function (e.g., genetic code, DNA replication and repair, etc.) of genes. During this period, investigators elucidated the molecular mechanisms leading to base pair substitution and frameshift mutations. At an important Bar Harbor Symposium in 1968, the influential human geneticist, James Crow, proposed that chemicals producing hereditary toxicity should be added to the list of adverse toxic endpoints to be regulated (34). In the following year, the newly formed EMS under the direction of Alexander Hollaender helped push this concept into the scientific spotlight, and lists of human diseases believed to be the consequence of single gene mutations (35) were viewed with a new perspective. Might the increasing exposure of humans to industrial chemicals increase the frequency of such mutations?

The specific health risk from mutations in human genes is determined by evaluation of the site of damage. Mutations in germ cells (ova or sperm), or tissues that give rise to germ cells, will almost exclusively produce effects in the next (F1) or later generations (e.g., muscular dystrophy, hemophilia, sickle cell disease, etc.). Mutations in somatic cells produce effects in individuals of the exposed population (e.g., tumor initiation, atherosclerosis, etc.). The dual modes of action involved in the genetic damage directly influenced how genetic toxicity became integrated into regulatory strategies. Dual-track, hierarchical testing schemes were designed to identify somatic and/or germ cell hazard and risk. To date, greater progress in regulatory decision making has been made with testing for somatic cell diseases. The logistics and resources needed for assessing germ cell damage and risk are considerably more challenging than those needed for detecting somatic cell effects (36). Attempts to use in vivo somatic cell genotoxicity data to predict germ cell responses have generally produced results that were considered inadequate for reliable classification of germ cell risk (37,38). Appropriate assessment of mammalian germ cell effects require in vivo models with appropriate target tissue metabolism and bioavailability as well as the ability to screen thousands of offspring for the presence of newly induced mutations. The cost of an assessment of a single chemical for germ cell gene mutation in mice would equal or exceed the cost of conducting a cancer bioassay in the same species.

Even if one wished to conduct a germ cell mutation assessment in rodent models, critical scientific questions remain regarding our ability to interpret those results as they might apply to human germ cell risk.

- It is known that the human gene pool already carries a mutational load of about 5% that is passed from one generation to the next (39). What geneticists would like to know, but do not, is the origin of these mutations. Are they the result of accumulated spontaneous mutations or the result of the activity of exogenous mutagens (e.g., radiation, combustion of fossil fuels, etc.)?
- Another unanswered question is "what evidence exists indicating that exposure to man-made genotoxins is capable of adding to the mutational load in human germ cells and, if added, will those mutations add incrementally to societal health care costs?"
- Scientists and regulators also need to know if there are mutagens that might be target site–specific for germ cells such that they would be missed using tests designed only to detect damage in primary somatic cell cultures or continuous cell lines.
- The final question is "is there sufficient mechanistic information about mutation induction in mammalian species to be able to make accurate extrapolations from germ cell damage and risk in rodent models to germ cells in human populations?"

Attempts to provide answers to these questions have been only partially satisfactory. From the existing human mutational load, we know that healthcare burden to society is substantial, and new information developed from the Human Genome Project continues to add to the body of evidence linking mutation and adverse health effects and abnormalities in humans.

The goals of genetic toxicology testing are to identify and control or eliminate exposure to mutagens that could add additional mutations to the human genome of both somatic and germ cells. However, until genetic toxicologists can develop information addressing the questions raised in the previous list, it will be difficult for regulators to demand genetic assessment of mammalian germ cells on a routine basis. Alternatively, there are several in vitro and in vivo somatic cell assays measuring genotoxicity (40,21).

Estimates of health-related and medical costs to society from new somatic cell mutations are relatively straightforward. The elimination of, or reduction in, somatic cell exposure to mutagens should be followed by a concomitant reduction in somatic cell–based diseases. Unfortunately, the same outcome cannot be assumed for new transmissable mutations added to the human genome. New germ cell mutations can be fixed in the genome and over time distributed across human populations generation after generation (e.g., sickle cell disease and Lesch Nyhan syndrome). Once fixed in the genome, the mutations become part of the genetic load and elimination of the agent(s) responsible for induction of new mutations will not eliminate or reduce the incremental disease burden produced by them.

Because genetic toxicologists lack adequate answers for the questions listed above, application of genetic testing to current safety testing practices is heavily focused on developing information regarding damage to somatic cells. Expectations that information from somatic cell assays will provide risk management decisions simultaneously offering protection for germ cells rest with assumptions that (*i*) there are no unique germ cell mutagens that could go undetected in somatic cell assays, and (*ii*) bioavailability barriers in gonadal tissue prevent genetic damage, compared with somatic tissue, on a unit-dose basis (41,19). If these two assumptions are true, results from somatic cell assays should be biased toward overpredicting hazard to germ cells.

Use of Test Battery Data to Reach Regulatory Classifications or Decisions

The strategy of combining assays into test batteries is common to several areas of safety testing today. However, construction and use of test batteries for safety evaluation appear to be first used in the field of genetic toxicology. The primary reason underlying development of batteries in genetic toxicology is directly linked to the multitude of modes of action through which gene structure or function can be altered along the genotoxicity pathway shown in Figure 1 (37). As illustrated in the figure, three specific endpoints of the pathway have been shown to produce genetic alterations directly leading to adverse health risks in humans, namely gene mutation (six), clastogenicity (five), and aneuploidy (seven). Since no single short-term test

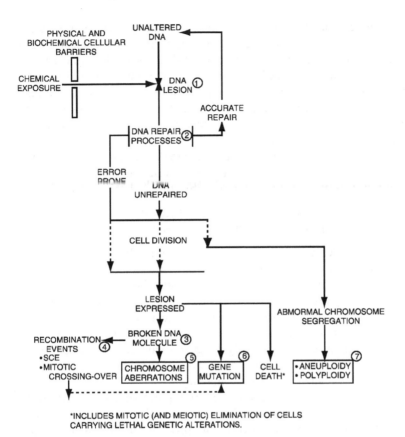

Figure 1 Hypothetical pathway from environmental exposure to the production of stable gene and chromosome mutations in eukaryotic cells. Steps 1–7 identify points where common genetic toxicology assays to measure hazard have been developed.

is able to provide simultaneous detection of all three endpoints, combination of test methods have been developed. The simplest battery would include a set of tests that detect each of the three endpoints. Tests which measure endpoints one through four are often included in test batteries but generally carry less weight because they are one or two steps removed from the primary DNA lesion that is the cause of the disease. Current guidelines require test batteries with tests for chromosome aberrations and gene mutation.

Recently, some regulatory agencies are considering the addition of tests specifically designed to detect aneuploidy as this endpoint is not covered by all guidelines, particularly ICH (42).

The battery testing strategy, perfected in genetic toxicology, has migrated to other areas of toxicology, most notably immunotoxicology, in vitro neuropharmacology, and in vitro target organ toxicity testing (43).

The number and diversity of genetic toxicology testing batteries are too large to compile and summarize. Among the earliest published strategies for genetic toxicology testing were those published by Bridges and Flamm (6,7). These strategies addressed the protection of both germ and somatic cell populations. Following numerous publications that reported the high correlation between rodent carcinogenesis and positive responses in short-term genotoxicity tests, batteries were rapidly reoriented toward the detection of rodent carcinogens (12,44).

Establishing the battery testing strategy created a unique set of dilemmas for those who planned to employ test batteries to assess and render decisions about safety. The following describes three of these obstacles.

The Need for Standardized Processes to Evaluate Heterogeneous Results from Multiple Test Batteries

The nature of short-term genetic tests is that they render dichotomous (+ or −) results. Unless the results within the array of tests are consistent (all + or all −), one must develop a set of rules to interpret a heterogeneous mixture of positives and negatives. At the outset of battery development, there was no consensus among genetic toxicologists on the rules for interpretation. Therefore, it was not uncommon for the same agent to be classified differently using two different batteries supposedly designed to identify the same hazard. This resulted in confusion and a lack of confidence that genetic testing could be reliably used for regulatory decision making.

A weight-of-evidence process for combining heterogeneous data (e.g., EPA's genetic risk assessment guidelines) is becoming accepted by many regulatory and advisory organizations. The process outlined in the mutagenicity risk assessment guidelines generated a qualitative classification, determined by the numbers and types of positive results, similar to those developed for carcinogenesis (e.g., limited, suggestive, and sufficient). This approach appeared likely to facilitate the interpretation of genetic toxicology batteries with heterogeneous results, but controversies arose regarding what weight (relevance) specific tests should contribute to the weight-of-evidence (e.g., should in vivo test results contribute more weight than in vitro test results?). Most of these questions were not easily resolved.

A committee created under the auspices of ICPEMC worked for several years to develop a rule-based semiquantitative approach for the assessment and interpretation of results from genetic testing batteries composed of both in vitro and in vivo assays (45). The system included options for weighting the contribution of tests according to their performance and combined results using a systematic method. The output of the system is a weight-of-evidence score that can be used to classify and rank chemicals based on heterogeneous data.

A secondary finding derived from the ICPEMC committee activities was that the universe of chemicals cannot be divided into two groups: one group that is genotoxic and the other that is not (46). The data published for the 113 chemicals included in the first ICPEMC database as well as for the chemicals added in the update show that, if tested exhaustively, any chemical will produce a positive result in some test for DNA damage. Consequently, quantitation of the weight-of-evidence process becomes important for developing a consistent measure of genotoxic hazard potential.

While there are highly refined and reasonably well-validated systems, such as the ICPEMC approach, for the most part, interpretation of genetic test battery results today continues to rely upon the subjective judgement of individual experts.

The Need to Extrapolate Findings from In Vitro to In Vivo Testing Models

Extrapolation of data between species, tissues, and treatment methods is essential for use in nonhuman safety testing. Although some multitest batteries that were developed consisted of only in vitro or only in vivo assays, most batteries, particularly for genetic toxicology assessments, employ both formats either concomitantly or sequentially. The question that continually emerges from such batteries is "what is the appropriate methodology for interpreting conflicting test results between cell-based systems and results from intact animals?" The conventional thinking was that animal-based results, particularly mammalian species, would be more reliable and relevant than cell-based responses, and should therefore take precedence over noncorresponding in vitro responses. Justification for this view is tied to the relevance of the in vivo model's bioavailability and metabolism. It has also been demonstrated that many in vitro tests are susceptible to false positive responses because of excessive cytotoxicity or other treatment environment extremes (47–49).

Arguments against the ability to override positive in vitro results with negative in vivo responses are generally constructed around the low sensitivity intrinsic to most in vivo assays. The argument offered is that protocol designs used in vivo will only detect relatively potent genotoxins and negative results should not negate positive in vitro tests measuring the same endpoint. The debate around such logic continues and no consensus regarding an answer has emerged.

The Need for Validation Methods

Conventional toxicology tests are intrinsically "self-validating" in that the animal model expresses the adverse effect in question. Short-term tests are generally based on concordance and attempt to use surrogate endpoints to predict those expressed in humans. Consequently, they require formal validation. Much has been written about test validation, but like the previous two issues, very little consensus has been established. Proper validation is essential if genetic toxicology test results are going to be used to predict the occurrence of a different type of toxic endpoint. In genetic toxicology, an example of this is the use of Ames bacterial mutation test results to predict rodent carcinogenesis. The most straightforward approach to validate this use of a short-term test is to calculate concordance between the responses of the two systems for a specific set of test materials (50). For purposes of screening large sets of chemicals or test samples, concordance assessments may be appropriate; however, a more rigorous process of validation would be needed if the Ames test were to be proposed as an alternative to the rodent bioassay for classifying chemicals as carcinogenic risks to humans. None of the currently used short-term tests for mutation has been validated to that extent (although the p53 transgenic mouse model, which essentially measures mutation, is currently used in limited situations as an alternative to the 104 week mouse cancer bioassay).

Difficulties that confound test or test battery validation begin with different opinions around the definition of the term "validation." In seeking to develop a somewhat universally accepted definition of validation, several organizations established committees and working groups to address the issue (43). One of the more comprehensive treatments of test system validation was developed by an international workshop on test system validation organized in 1990 by the Johns Hopkins Center for Alternatives to Animal Testing and the European Research Group for Alternatives in Toxicity Testing. The purpose of the workshop was to develop guidance for validation of predictive alternative tests, including genetic toxicology. The definition developed from this workshop appeared to be relatively simple; validation is the "process whereby the reliability and relevance of a method are assessed for a specific purpose" (51). The implications of this definition to the practicality surrounding the validation of single or batteries of tests are substantial. Normally, in the first step of validation, one must decide on the purpose of the test or the test battery. All too often, this step is not conducted with sufficient clarity. The consequence is that the design of the validation process for the test or battery turns out to be inadequate. The other two elements of validation are equally critical. The first is reliability; it establishes the precision and reproducibility of a method. In assessing reliability, one must decide on acceptable levels for reliability (e.g., type 1 error rate) that assures the same precision as that of the method it is attempting to replicate. It is possible that a test or battery of tests may be very precise but is not relevant for its intended purpose. For example, the conventional bacterial mutation assay developed by Bruce Ames might be proposed as an alternative to rodent cancer bioassays (i.e., its purpose). The test may be shown to have adequate reliability to identify bacterial mutagens, but because some cancers do result directly from mutation, validation of the test fails on the basis of relevance.

Other details that must be considered in developing a reliable validation process consist of knowing the number of agents and chemical classes that must be included in the validation set to assure the appropriate level of accuracy. The same process and requirements would hold true for the validation of test batteries (51).

Validation programs for genetic tests and test batteries, as alternatives for the rodent cancer bioassay, have often been performed without adequate attention to their reliability and/or relevance. As a result, the influence of these methods on regulatory decisions regarding chemical or drug safety has often been inconsistent and remains controversial (30).

INFLUENCE OF GENETIC TOXICOLOGY RESEARCH ON REGULATORY POLICY

Because of the existing difficulties in establishing a cause and effect relationship between mutation induction in mammalian germ cells and the production of specific health effects, no materials have been regulated based solely on their mutagenicity. Today, genetic toxicology data contribute to toxicology profiles used in the approval processes for human and animal medicines/devices, food additives, and regulated environmental chemicals, and are found in labeling information attached to pharmaceutical products. Depending on the type of product, its intended use, and level of human exposure, genetic toxicology data play different roles:

- The data may be part of a rodent bioassay trigger process for environmental materials that would not automatically be required to have chronic rodent carcinogenicity tests conducted as part of their approval process (e.g., EPA Toxic Substances Control Act).
- Genetic toxicology test results have been used in assessing mode of action for rodent carcinogens. Under various risk assessment guidelines, DNA reactive (genotoxic) carcinogens are treated differently from non-DNA reactive (epigenetic) carcinogens with respect to risk assessment modeling. Cancer risk assessments for DNA reactive carcinogens generally employ no threshold models such as the linearized multistage model (52).
- The data may be used as part of a safety assessment database to qualify a material for limited human exposure before the completion of rodent carcinogenicity studies (e.g., Food and Drug Administration drug approval process for Phase I clinical trials). If an adequate battery of tests assessing genotoxicity shows no evidence for DNA reactivity and/or damage, regulatory bodies use this data as a "comfort factor" before proceeding to human exposures. Conversely, strong positive responses in one or more tests of the battery may delay human clinical studies until adequate assurances of noncarcinogenicity are provided.

In addition to these direct applications to regulatory decision-making processes, data from genetic toxicology assays are used in other, indirect, ways by regulatory and/or advisory bodies.

Molecular epidemiology: Several, mildly invasive, measurements of DNA damage can be performed on human populations. The most frequently used methods are (*i*) formation of DNA adducts in isolated cells, (*ii*) induction of sister chromatid exchange or chromosome breakage in peripheral blood cultures, and (*iii*) production of single or double strand DNA breaks in tissue biopsies using the comet assay (53,54). These tests are predictive for DNA reactive carcinogens, and data derived from human monitoring studies are valuable in establishing or confirming actual human contact (qualitative or quantitative) to the test agent under relevant exposure conditions (48). Using some of the above technologies to monitor populations for DNA reaction products can be extremely valuable in confirming a variety of risk assumptions such as dose rate kinetics, cumulative effects, and the range of human susceptibility.

Assessment of low dose effects: Understanding the dose rate kinetics for DNA lesions at very low doses will aid in the development of improved risk assessment modeling. Investigations of DNA damage conducted with low doses of radiation and genotoxic carcinogens demonstrate that a multitude of mechanisms abolish linearity of the dose–response kinetics. Phenomena such as DNA repair induction, distribution of DNA damage, requirement for multiple hits for production of chromosome breaks, and programmed cell death dampen linearity at very low doses and should eventually be built into risk assessment models for DNA reactive carcinogens.

LOOKING INTO THE FUTURE OF GENETIC TOXICOLOGY'S ROLE IN SAFETY TESTING STRATEGIES

The practices currently embodied in safety testing are slowly evolving as a result of technology advances such as toxicogenomics, proteomics, high-resolution chemical detection techniques, and increases in our understanding of cell-signaling processes. Aspects of these new technologies will also influence the detection of mutagens and clastogens.

The introduction of change to regulatory practice and procedures is typically a very slow process; however, today's regulatory testing for genetic toxicology employs tests that are between 25 and 30 years old (e.g., the initial Ames test methods were published in 1975). With increased knowledge of the types of genetic changes and the instability associated with diseases such as cancer, limitations associated with the incumbent assays have been identified and discussed extensively (30). Attempts to use the existing battery of tests to look for aneugens is a positive step, but scientists believe other DNA alterations may possibly play a role in cancer initiation (e.g., DNA methylation inhibition of gene expression) and will not be identified using the current test batteries. New methods available to monitor a wider range of DNA damage such as in vivo transgenic mice for mutation assessment and cell transformation (the Syrian hamster embryo cell assay) are now reliable enough to provide more relevant information regarding risk of genotoxic effects that could result in somatic cell diseases (55). Some of these new methods allow investigators to detect a broader, and possibly more relevant, range of genetic alterations. However, they may not have been accepted more widely because of special expertise required to carry out the tests and higher costs than incumbent assays. As discussed earlier, there is also the enormous issue of assay validation. For a new, and possibly better, test to take the place of one of the current ICH battery tests, it would have to bring with it a very large experience base developed with multiple chemical classes. Developing such a database for a relatively new method requires both considerable time and money. Few organizations have the resources available to commercialize (fully validate) a new method. In the short term, the relevance and real value of data from genetic testing could be improved by some of these methods if the issues of higher cost and validation could be overcome. The prospects for improved alternative tests to replace the current basic assays accepted by ICH or EPA are extremely low.

The most likely scenario leading to genetic toxicology testing in the future will be tied to more universal paradigm shifts in biotechnology. The new technology will be looking directly at the cause of toxicity and will be virtually "self-validating." Examples of candidates for shifts of this type may well be tied to information developing from the Human Genome Project. As information about the role of gene expression in normal and genetically altered cells becomes more readily available, toxicogenomic and proteomic profiles should be able to pinpoint dangerous chemicals, and these technologies can potentially also identify the lowest exposures that genetically alter cells in the most susceptible tissues. While toxicogenomic data cannot identify mutations per se, gene expression patterns showing overexpression of genes involved in DNA repair or apoptosis are relevant surrogates and probably more reliable indicators of real genotoxic hazard and they can be performed directly on the animals expressing the toxic endpoint.

SUMMARY

The discipline of Genetic Toxicology is still trying to find a comfortable home in the regulatory arena. While there is no controversy regarding the importance of the genome in determining an organism's health and well-being, scientists and regulators have not reached a consensus on how best to use the results of today's technologies in product approval or human risk estimates. The widest application currently is to use genetic toxicity data as an early signal for mutagenic carcinogens. With the development of new, more relevant, technologies there is hope that monitoring for integrity of the genome will become simpler to integrate into formal safety assessments (i.e., regulatory decisions based on risk of germ cell mutation will become widely accepted).

Advancements in our understanding of the human genome and of how genetic changes affect human health will eventually lead to better tests for genetic damage and to significantly improved methods for predicting and preventing adverse health effects associated with new product development.

REFERENCES

1. Douglas G. Discussion Forum: the inclusion of an assay for inherited congenital malformations in the assessment of mutagenicity. Mutagenesis 1990; 5(5): 421–423.
2. Lohman P, Mendelsohn M, Moore D II, et al. A method for comparing and combining short-term genotoxicity test data: the basic system. Mutat Res 1992; 266:7–25.
3. Lohman P, Morolli B, Darroudi F, et al. Contributions from molecular/biochemical approaches in epidemiology to cancer risk assessment and prevention. Environ Health Perspec 1992; 98:155–165.
4. Brusick D. Current trends in the development of short-term predictive toxicology assays. Concepts Toxicol 1984; 1:190–199.
5. Ray V. Application of microbial and mammalian cells to the assessment of mutagenicity. Pharmacol Rev 1979; 39(4):537–546.
6. Bridges B. Some general principles of mutagenicity screening and a possible framework for testing procedures. Environ Health Perspect 1973; 6:221–227.
7. Flamm W. A tier system approach to mutagen testing. Mutat Res 1974; 26: 329–333.
8. Lee W, Abrahamson S, Valencia R, von Halle E, Wurgler F, Zimmering S. The sex-linked recessive lethal test for mutagenesis in *Drosophila melanogaster*. Mutat Res 1983; 123:183–279.
9. Ehling U. Methods to estimate the genetic risk. In: Obe G, ed. Mutations in Man. Heidelberg, Berlin: Springer-Verlag, 1984:292–318.
10. Ames B, Durston W, Yamaski E, Lee F. Carcinogens are mutagens: a simple test system combining liver homogenates for activation and bacteria for detection. Proc Natl Acad Sci USA 1973; 70:2281–2285.
11. Waters M, Stack H, Brady A, Lohman P, Haroun L, Vainio H. Use of computerized data listings and activity profiles of genetic and related effects in the review of 195 compounds. Mutat Res 1988; 205:294–312.
12. Purchase I, Longstaff E, Ashby J, et al. Evaluation of six short-term tests for detecting organic chemical carcinogens and recommendations for their use. Nature 1976; 264:624–629.
13. Ames B. Identifying environmental chemicals causing mutations and cancer. Science 1979; 204:587–593.
14. Weisburger J, Williams G. Carcinogen testing: current problems and new approaches. Science 1981; 214:401–407.
15. Buzard G. Studies of oncogene activation and tumor suppressor gene inactivation in normal and neoplastic rodent tissue. Mutat Res 1996; 365:43–58.
16. Committee 17 Report. Environmental mutagenic hazards. Science 1975; 187: 503–517.
17. Ehling U. Germ-cell mutations in mice: standards for protecting the human genome. Mutat Res 1989; 212:43–53.
18. Sobels F. Models and assumptions underlying genetic risk assessment. Mutat Res 1989; 212:77–89.

19. Favor J, Layton D, Sega G, et al. Genetic risk extrapolation from animal data to human disease: an ICPEMC taskgroup report. Mutat Res 1995; 330:23–34.
20. National Center for Toxicological Research. Short term in vitro testing for carcinogenesis, mutagenesis and toxicity. In: Berky J, Sherrod C, eds. Collected Proceedings of the First and Second Working Conference of Toxicity Testing In Vitro 1975/1976. Philadelphia, PA: The Franklin Institute Press, 1978.
21. Brusick D, Auletta A. Developmental status of bioassays in genetic toxicology: A Report of Phase II of the U.S. Environmental Protection Agency Gene-Tox Program. Mutat Res 1985; 153:1–10.
22. Russell L, Aaron C, de Serres F, et al. Evaluation of mutagenicity assays for purposes of genetic risk assessment: A Report of the U.S. Environmental Protection Agency Gene-Tox Program. Mutat Res 1984; 134:145–157.
23. Garrett N, Stack F, Gross M, Waters M. An analysis of the spectra of genetic activity produced by known or suspected human carcinogens. Mutat Res 1984; 134:89–111.
24. Ray V, Kier L, Kannan K, et al. An approach to identifying specialized batteries of bioassays for specific classes of chemicals: class analysis using mutagenicity and carcinogenicity relationships and phylogenetic concordance and discordance patterns. 1. Composition and analysis of the overall data base. Mutat Res 1987; 185(3):197–241.
25. US Environmental Protection Agency (USEPA). Guidelines for carcinogen risk assessment. Fed Reg 1986a; 51:33992–34003.
26. US Environmental Protection Agency (USEPA). Guidelines for mutagenicity risk assessment. Fed Reg 1986b; 51(185):34006–34012.
27. International Commission for Protection Against Environmental Mutagens and Carcinogens (ICPEMC). Screening strategy for chemicals that are potential germ-cell mutagens in mammals. Committee 1 Final Report. Mutat Res 1983a; 114:117–177.
28. Tennant R, Margolin B, Shelby M, et al. Prediction of chemical carcinogenicity in rodents from in vitro genetic toxicity assays. Science 1987; 236:933–941.
29. Haseman J, Zeiger E, Shelby M, Margolin B, Tennant R. Predicting rodent carcinogenicity from four in vitro genetic toxicity assays: an evaluation of 114 chemicals studied by the National Toxicology Program. J Am Stat Assoc 1990; 85(412):964–971.
30. Zeiger E. Identification of rodent carcinogens and noncarcinogens using genetic toxicity tests: premises, promises, and performance. Regul Toxicol Pharmacol 1998; 28:85–95.
31. Auletta A, Dearfield K, Cimino M. Mutagenicity test schemes and guidelines: U.S. EPA office of pollution prevention and toxics and office of pesticide programs. Environ Mol Mutagen 1993; 21:38–45.
32. Kirkland D. Genetic toxicology testing requirements: official and unofficial views from Europe. Environ Mol Mutagen 1993; 21:8–14.
33. OECD. OECD guideline for the testing of chemicals: No. 471, bacterial reverse mutation test; No. 473, in vitro mammalian chromosome aberration test, No. 474, mammalian erythrocyte micronucleus test; No. 476, in vitro mammalian cell gene mutation test. Organization for Economic Cooperation and Development, Paris, 1997.

34. Crow J. Chemical risk to future generations. Scientist Citizen 1968; 10:13–117.
35. McKusick VA. Mendelian Inheritance in Man: Catalogs of Autosomal Dominant, Autosomal Recessive and X-liked Phenotypes. 5th ed. Baltimore: Johns Hopkins University Press, 1978.
36. International Commission for Protection Against Environmental Mutagens and Carcinogens (ICPEMC). Regulatory approaches to the control of environmental mutagens and carcinogens. Committee 3 Final Report. Mutat Res 1983b; 114:179–216.
37. International Commission for Protection Against Environmental Mutagens and Carcinogens (ICPEMC). Estimation of genetic risks and increased incidence of genetic disease due to environmental mutagens. Committee 4 Final Report. Mutat Res 1983c; 115:255–291.
38. Waters M, Stack H, Jackson M, Bridges B, Adler I. The performance of short-term tests in identifying potential germ cell mutagens: A qualitative and quantitative analysis. Mutat Res 1994; 341:109–131.
39. Brusick D. Evolution of testing strategies for genetic toxicity. Mutat Res 1988; 205:69–78.
40. Albertini R, Nicklas J, O'Neill P. Somatic cell gene mutations in humans: biomarkers for genotoxicity. Environ Health Perspec Suppl 1993; 101(suppl 3): 193–201.
41. Adler I, Ashby J. The present lack of evidence for unique rodent germ-cell mutagens. Mutat Res 1989; 212:55–66.
42. Parry E, Henderson L, MacKay J. Guidelines for the testing of chemicals. Procedures for the detection of chemically induced aneuploidy: recommendations of a U.K. environmental mutagen society working group. Mutagenesis 1995; 10:1–14.
43. Goldberg A, Zurlo J, Epstein L. Validation revisited. In: van Zutphen L, Balls M, eds. Animal Alternatives, Welfare and Ethics. The Netherlands: Elsevier, 1997:1077–1082.
44. International Commission for Protection Against Environmental Mutagens and Carcinogens (ICPEMC). Mutagenesis testing as an approach to carcinogenesis. Mutat Res 1982; 99:73–91.
45. Brusick D, Ashby J, de Serres F, et al. A method for combining and comparing short-term genotoxicity test data: preface. A Report from ICPEMC Committee 1. Mutat Res 1992; 266:1–6.
46. Mendelsohn M, Moore D II, Lohman P. A method for comparing and combining short-term genotoxicity test data: results and interpretation. Mutat Res 1992; 266:43–60.
47. Müller L, Sofuni T. Appropriate levels of cytotoxicity for genotoxicity tests using mammalian cells in vitro. Environ Mol Mutagen 2000; 35:202–205.
48. Galloway S. Cytotoxicity and chromosome aberrations in vitro: experience in industry and the case for an upper limit on toxicity in the aberration assay. Environ Mol Mutagen 2000; 35:191–201.
49. Scott D, Galloway S, Marshall R, et al. Genotoxicity under extreme culture conditions. Mutat Res 1991; 257:147–204.

50. Ames B, McCann J, Yamasaki E. Methods for detecting carcinogens and mutagens with salmonella/mammalian microsome mutagenicity test. Mutat Res 1975; 31:347–364.
51. Balls M, Blaauboer B, Brusick D, et al. Report and recommendations of the CAAT/ERGATT workshop on the validation of toxicity test procedures. Alta 1990; 18:313–337.
52. Butterworth B, Conolly R, Morgan K. A strategy for establishing mode of action of chemical carcinogens as a guide for approaches to risk assessments. Cancer Lett 1995; 93:129–146.
53. Uziel M, Munro N, Katz S, et al. DNA adduct formation by 12 chemicals with populations potentially suitable for molecular epidemiological studies. Mutat Res 1992; 277:35–90.
54. Galloway S, Berry P, Nichols W, et al. Chromosome aberrations in individuals occupationally exposed to ethylene oxide, and in a large control population. Mutat Res 1986; 170:55–74.
55. Brusick D. Genetic toxicology. In: Hayes W, ed. Principles and Methods of Toxicology. 4th ed. Philadelphia: Taylor & Francis, 2001:819–852.

Index

Milton Keynes UK
Ingram Content Group UK Ltd.
UKHW020022071024
449327UK00032B/2886

9 780367 391089